Mobility-based Time References for Wireless Sensor Networks

ANALOG CIRCUITS AND SIGNAL PROCESSING

Series Editors:

Mohammed Ismail, The *Ohio State University*
Mohamad Sawan, *École Polytechnique de Montréal*

For further volumes:
http://www.springer.com/series/7381

Fabio Sebastiano • Lucien J. Breems
Kofi A.A. Makinwa

Mobility-based Time References for Wireless Sensor Networks

 Springer

Fabio Sebastiano
NXP Semiconductors
High Tech Campus 32
Eindhoven
Netherlands

Lucien J. Breems
NXP Semiconductors
High Tech Campus 32
Eindhoven
Netherlands

Kofi A.A. Makinwa
Delft University of Technology
Mekelweg 4
Delft
Netherlands

ISBN 978-1-4899-8828-7   ISBN 978-1-4614-3483-2 (eBook)
DOI 10.1007/978-1-4614-3483-2
Springer New York Heidelberg Dordrecht London

Printed on acid-free paper

Springer is part of Springer Science+Business Media (www.springer.com)

Preface

Measuring time is an ancient problem. Along history, man has developed more and more advanced timekeeping devices to keep up with his growing needs: mechanical clocks to increase accuracy; quartz watches to cut down on size and cost; atomic clocks to reach higher accuracy. Today, each node in a wireless sensor network (WSN) needs a time reference with even novel requirements: it must be cheap, of ultralow power, and fully integrated on a silicon die, thus not employing any off-chip component.

This book approaches the challenge of designing such reference in a systematic way. First, the reader is guided through the architectural choices and the trade-offs in the design of a wireless sensor node in order to understand the application of the time reference and its specifications. Based on that and on the review of several alternatives, mobility-based references are chosen as the preferred candidates. In the following parts of the book, they are thoroughly dissected by illustrating concepts, circuit design, temperature compensation schemes, and experimental validations, so that their potentiality as time references for WSN can be finally assessed.

The research presented in this book has been completed during my Ph.D. and I wish to thank those who helped me during that period: the European Commission for financial support in the Marie Curie project TRANDSSAT; the TRANDSSAT team, comprising Salvo Drago, Lucien Breem, Domine Leenaerts, Kofi Makinwa, and Bram Nauta; Onur Kaya and Rabindra Rijal for helping with the 0.16-μm-CMOS measurements; Maristella Spella and Muhammed Bolatkale for proofreading the manuscript; the Mixed Signals Circuits and Systems group of NXP Research, where the research was carried out; the Electronic Instrumentation Laboratory of Delft University. Finally, nothing would have been possible without the loving support of my family and my fiancée Elisa.

Eindhoven   Fabio Sebastiano

Contents

Chapter 1
Introduction

If the doors of perception were cleansed every thing would appear to man as it is, infinite.

William Blake
The Marriage of Heaven and Hell

1.1   Enhancing Perception

In the evening, a man comes back home. As he steps into the house, the lights turn on in the living room. In the kitchen, the sunlight is still strong enough and the lamps stay off. While he is browsing through his mail, the faint click of the windows upstairs is heard. The air in the bedroom has just reached the optimal temperature and it would get too cold now that the sun has set. He moves to the kitchen and reads on the fridge display: the milk inside has gone beyond its expiration date but its organoleptic and nutritional properties are still unaffected. Instead, it is time for the fruit in the bowl on the counter to get replaced, the display shows. At this point, he could recall on the display a variety of data: the temperature and humidity of each room; the pressure of his car's tyres; the status of the wine ageing in the cellar; whether water or any fertilizer is required in the garden; even the structural integrity of the house itself.

What is the impact on the man's daily life of having access to such amount of information? First, some common operations can be maximally optimized. The temperature and light of indoor environments can be set by exploiting solar exposure and external temperature variations, adapting them only when human presence is detected, and, thus, saving energy [1]. The end of the lifetime of food and other consumable or perishable goods (such as tyres, mechanical parts, even constructions ...) can be accurately detected, ensuring safety and maximum utilization [2, 3]. Moreover, completely new things are possible, such as continuous monitoring of biomedical parameters (heart and respiratory rate, oxygen in the blood, ...) and telemedicine [4, 5]. Possibilities are limitless.

F. Sebastiano et al., *Mobility-based Time References for Wireless Sensor Networks*,
Analog Circuits and Signal Processing, DOI 10.1007/978-1-4614-3483-2_1,
© Springer Science+Business Media New York 2013

This vision can become true if our surroundings (our home, our workplace, the streets) would be populated by a large amount of tiny sensors [6]. Temperature and humidity sensors would be placed in any corners of buildings, chemical sensors in any food container, stress sensors and accelerometers on any mechanical or structural part, even inside walls or on the wings of an airplane. Those sensors must be small and cheap enough to be placed in any milk carton without being noticed. Each sensor node would acquire at least a physical parameter and transmit it to the other nodes. The data would start bouncing from one node to the other, so that it would be available in any point of such network and thus be accessible everywhere. Those *Wireless Sensor Networks (WSNs)* would be an invisible infrastructure surrounding us and enhancing our own senses and thus our perception of the physical world.

1.2 References for Wireless Communication

Wireless communication is unavoidable for a pervasive and ubiquitous technology as WSN. Since power can not be wired to the sensor node, energy is provided by a small battery or is harvested from the environment, for example from solar light or mechanical vibrations. Such small amounts of energy must be carefully economized to ensure proper functionality for a long time. The average power consumption of a WSN node must then be extremely low, in the order of $100\,\mu W$ [7]. The wireless interface, practically implemented with a radio, must be obviously cheaper and smaller than the whole node. Thus, a natural choice would be to build it as a fully integrated microelectronic system. However, since the 1920s, any radio has been equipped with at least one bulky external component, i.e. a quartz crystal [8]. The crystal is used as a piezoelectric resonator: the high accuracy of its resonant frequency is exploited to get an accurate time and frequency reference, which is necessary in any practical transceiver.

A frequency reference is needed to ensure that the transmitted signal lies in the right portion of the RF spectrum. Apart from the practical need to respect spectral regulations, the transmitter and the receiver must be tuned to each other, i.e. use the same frequency channel, and thus they must share a common reference.

A time reference is needed when a wireless node is receiving. The node could wait for an incoming message by continuously listening to the channel. In terms of energy, this would be very inefficient in a network where messages are sporadically sent. This is the case for systems where the parameters of interest (the temperature in a room, the pressure of a tyre) are changing very slowly. For the receiving node, it would be helpful to know when a message is expected, so that it could turn on only when needed. This can be accomplished with the combination of a synchronization algorithm and a time reference.

However, external components must be avoided in small low-cost WSN nodes. A fully integrated low-power alternative to the crystal-based time/frequency reference must then be found.

1.3 Fully-Integrated Time References

Time/frequency references can be based on several operating principles. For example, crystal oscillators generate a frequency locked to the mechanical resonant frequency of a quartz crystal. Other physical principles more amenable to silicon integration includes, among others: locking to the thermal delay of silicon [9]; locking to the resonance of an LC network [10]; locking to the mechanical resonance of micromachined integrated structures, i.e. Microelectromechanical System (MEMS) [11]. Some of those references achieve accuracies comparable to that of crystal oscillators. However, some, such as LC-based and thermal-diffusivity-based references, require high power consumption (in the order of a few milliwatts), while others, such as MEMS-based references, requires non-standard silicon processing. The former can not be powered by the typical WSN energy sources, whereas the latter would be more expensive than nodes implemented in a standard IC process. Regarding the choice of the process, a baseline deep-submicron CMOS process is the preferred choice, because of the possibility to integrate the time reference together with the low-power blocks for the digital and RF processing needed on-board [12].

1.4 Motivation and Objectives

Despite many alternatives, no CMOS fully integrated time reference simultaneously offers high accuracy, low power consumption and low cost, and thus no ideal time reference for WSN is currently available. The main aim of this work is to describe a solution to this problem: how to realize a low-power low-cost CMOS time reference for WSN.

The research has been conducted by tackling the problem from two sides: at the system level and at the circuit and technology level. At the system level, the architecture of the WSN node and the network protocol ruling the interactions between the nodes can be analyzed. The objective is to find architectures and protocols that relax the accuracy requested to the time and frequency references mentioned above, while ensuring proper functionality within the tight power budget of the node. The main specifications of the references, i.e. accuracy and power consumption, can be derived from the choice of the system parameters, such as the synchronization strategy and the data modulation.

At the circuit level, the objective is to identify a physical principle that can provide the required accuracy within the power budget defined at the system level and to optimize its implementation for maximum accuracy. Particular care must be taken to ensure that the specified accuracy is met when subject to variations of Process, supply Voltage and Temperature (PVT). The fundamental and practical limitations of the selected reference must be studied both analytically and experimentally. The final objective of this book is then to describe the design of a fully integrated low-power CMOS time reference that is robust to PVT variations.

1.5   Organization of the Book

Chapter 2 is devoted to the analysis of time and frequency accuracy in a WSN node. The architecture of the node, the synchronization protocol in the network and the data modulation scheme are discussed and a solution is proposed to lower the accuracy required of the time reference to 1%.

Various existing fully integrated time references are reviewed in Chap. 3. Several approaches (including the use of RC and LC oscillators, thermal-diffusivity-based references and full-MOS references), their fundamental limitations and the practical limitations of their implementations are discussed. The shortcomings of their employment in a WSN node are underlined, showing why a different solution is sought.

In Chap. 4, a novel low-power time reference is proposed, which is referenced to the electron mobility in a MOS transistor. The impact of the different source of errors are treated, together with methods to reduce them. An implementation in a deep-submicron CMOS process (65-nm CMOS), suitable for WSN applications, is presented to experimentally prove the robustness to PVT variations. Additional designs in a more mature technology (0.16-μm CMOS) are described to demonstrate the portability of the proposed reference concept and its sensitivity to different processes and process options.

Chapter 5 deals with the compensation of the mobility-based reference for temperature variations, which are the main source of inaccuracy of this kind of reference. Due to the large temperature dependence of electron mobility, any adopted temperature compensation algorithm requires knowledge of the reference's temperature with high accuracy. A large part of the chapter is then devoted to the realization of a highly accurate low-power temperature sensor in deep-submicron CMOS and to its integration with the mobility-based reference.

Finally, Chap. 6 summarizes the main findings of this work and outlines possibilities for future research.

References

1. Guo W, Zhou M (2009) Technologies toward thermal comfort-based and energy-efficient HVAC systems: A review. In: IEEE International Conference on Systems, Man and Cybernetics, pp 3883–3888. DOI 10.1109/ICSMC.2009.5346631
2. Ergen S, Sangiovanni-Vincentelli A, Sun X, Tebano R, Alalusi S, Audisio G, Sabatini M (2009) The tire as an intelligent sensor. IEEE Trans Comp Aided Des Integr Circ Syst 28(7):941–955. DOI 10.1109/TCAD.2009.2022879

3. Harms T, Sedigh S, Bastianini F (2010) Structural health monitoring of bridges using wireless sensor networks. IEEE Instrum Meas Mag 13(6):14–18. DOI 10.1109/MIM.2010.5669608

4. Penders J, van de Molengraft J, Brown L, Grundlehner B, Gyselinckx B, Van Hoof C (2009) Potential and challenges of body area networks for personal health. In: Annual International Conference of the IEEE Engineering in Medicine and Biology Society, pp 6569–6572. DOI 10.1109/IEMBS.2009.5334004

5. Rashvand H, Salcedo V, Sanchez E, Iliescu D (2008) Ubiquitous wireless telemedicine. IET Comm 2(2):237–254. DOI 10.1049/iet-com:20070361

6. Weber W, Rabaey JM, Aarts E (eds) (2005) Ambient intelligence. Springer, New York

7. Rabaey J, Ammer M, da Silva J JL, Patel D, Roundy S (2000) Picoradio supports ad hoc ultra-low power wireless networking. Computer 33(7):42–48. DOI 10.1109/2.869369

8. Bottom V (1981) A history of the quartz crystal industry in the USA. In: 35th Annual Frequency Control Symposium, pp 3–12. DOI 10.1109/FREQ.1981.200452

9. Kashmiri SM, Pertijs MAP, Makinwa KAA (2010) A thermal-diffusivity-based frequency reference in standard CMOS with an absolute inaccuracy of $\pm 0.1\%$ from -55 to $125°C$. IEEE J Solid State Circ 45(12):2510–2520

10. McCorquodale M, Gupta B, Armstrong W, Beaudouin R, Carichner G, Chaudhari P, Fayyaz N, Gaskin N, Kuhn J, Linebarger D, Marsman E, O'Day J, Pernia S, Senderowicz D (2010) A silicon die as a frequency source. In: IEEE International Frequency Control Symposium, pp 103–108. DOI 10.1109/FREQ.2010.5556366

11. Nguyen CC (2007) MEMS technology for timing and frequency control. IEEE Trans Ultrason Ferroelect Freq Contr 54(2):251–270. DOI 10.1109/TUFFC.2007.240

12. Ammer J, Burghardt F, Lin E, Otis B, Shah R, Sheets M, Rabaey JM (2005) Ultra low-power integrated wireless nodes for sensor and actuator networks. In: Weber W, Rabaey JM, Aarts E (eds) Ambient intelligence. Springer, New York

Chapter 2
Fully Integrated Radios for Wireless Sensor Networks

2.1 Introduction

In the future, technology will be hidden in the environment and invisible to the user but, at the same time, responsive and adaptive to user interaction and environmental variations [1]. For example, smart buildings will become aware of the presence of people and of their needs: thanks to this, temperature and light conditions will be adapted automatically for the best comfort and optimal power consumption. The realization of such a vision requires a technology that can sense, process and respond to external stimuli, both human and environmental, coming from many spots of a large environment, such as a room, a house or a whole building. An answer to such needs may come from *WSNs*. These are networks that consist of a large number of energy-autonomous nodes deployed into the environment to collect physical data. Each node is equipped with sensors, digital and analog processing units and a radio transceiver [2]. Physical parameters, such as temperature, sound, light conditions, etc., are sensed and processed by each node. The resulting information is transmitted from node to node and propagates through the network, until it is collected by a central data sink or used by the network itself in distributed algorithms. Dense networks, composed of hundreds or thousands of devices, are required to accurately monitor an environment. Consequently, each node must be extremely cheap to limit the cost of the network and make this technology economically feasible. Moreover, the nodes must be small enough to be hidden, in order to be invisible to users and not affect the surrounding environment.

Early applications of sensor networks were mostly military. They ranged from very large, still human-controlled networks, such as acoustic sensors on the ocean bottom to track quiet Soviet submarines during the Cold War, to the more autonomous systems developed from the 1980s and 1990s, employed for instance for tracking aerial targets or for unattended ground monitoring [3]. The devices used in those systems were bulky and expensive, and only in the recent past have advances in sensing, processing and wireless technologies led to substantial reductions of both the cost and size of nodes for WSNs. Commercial companies, such as Crossbow,

F. Sebastiano et al., *Mobility-based Time References for Wireless Sensor Networks*,
Analog Circuits and Signal Processing, DOI 10.1007/978-1-4614-3483-2_2,
© Springer Science+Business Media New York 2013

Dust Networks and EnOcean, manufacture devices lighter than 100 g and smaller than a matchbox [4–6], which can address a wide range of civil and industrial applications: the control of heating, ventilation and air-conditioning (HVAC) systems in buildings, habitat monitoring, precision agriculture, the monitoring of vital signs and the control of food during processing and storage [7]. These nodes are generally battery powered. Since battery replacement is unpractical and expensive in dense networks, the lifetime of the network is severely limited by battery size and by the (poor) energy efficiency of the nodes. Even with small solar cells and other renewable sources of energy, the number of possible applications is limited by the energy available in the environment and by the power required by the electronics.

The next step leading to truly *disappearing electronics* [8] will be the integration of the whole node in silicon [9, 10]. CMOS technology is currently used to fabricate complex Systems-on-Chip (SoCs), comprising digital processors, RF front-ends, analog circuitry and even some sensors. Ideally, the antenna should be the only sexternal component,[1] since research indicates the feasibility of integrating energy collection and storage [12, 13]. However, one block is often forgotten: the frequency reference, which usually uses an external off-chip crystal. Unfortunately, integration of the reference usually implies a degradation of its performance, especially in terms of accuracy, as it will be shown in Chap. 3. The impact of such degradation on the design of a WSN node will be discussed in this chapter.

Apart from providing clock signals for digital circuits and some switched analog circuits, the reference has two main functions in a WSN node:

- A time reference[2] for synchronization of the node with other nodes in the network.
- A frequency reference which ensures that the right portion of the RF spectrum is used for communication.

After a more detailed description of the specifications of a node in Sect. 2.2, the requirements on timing accuracy for network synchronization and on the frequency accuracy for RF communication will be treated separately in Sects. 2.3 and 2.4, respectively. Finally, the proposed architecture of the node will be described in Sect. 2.5.

2.2 Requirements for a WSN Node

As mentioned in the previous section, WSN nodes must be cheap, small and extremely low power. Those specifications are described here in more detail, in order to derive a list of requirements (shown in Table 2.1) that will be used throughout this chapter.

---

[1]Depending on the frequency of operation and the required performance, antennas can also be integrated on silicon [11].

[2]In the following, the expressions "time reference" and "frequency reference" are interchangeable.

Table 2.1 WSN node requirements

Parameter	Value
Average power dissipation	$\leq 100\,\mu\text{W}$
External components	Only antenna and energy source
Number of nodes (n_{nodes})	100
Distance between two nodes (d)	$<10\,\text{m}$
Packet rate (PR)	1 pkt/min
Packet payload (N_{pl})	100
Bit error rate (BER)	10^{-3}

2.2.1 Application

Sensing of environmental parameters is a typical application for WSN. As an example, we can consider the monitoring of temperature, humidity and atmospheric pressure to optimally control the HVAC system in a building, or to collect scientific data or the ascertain the state of pollution in a natural habitat, such a forest. Focusing on this particular example, we can derive some general specifications for the WSN.

We consider a network containing a large number of low power wireless sensor nodes and a minimum of one high-power data sink node. Data are spread over the network through multi-hop routing and are collected in the data sink for further use by a central control system. We assume that the data sink is able to receive data in the WSN format and does not affect the power consumption of the network—for this reason it is not taken into account in the following. The network traffic is inherently low, since environmental parameters can be encoded in a few bytes and are generally varying slowly enough to be sensed at low rates. Moreover, if data redundancy (due for example to the spatial vicinity of nodes sampling the same parameter) is reduced by aggregating data, the number of transmissions can be decreased [14]. Requirements on the Quality of Service (QoS), such as a maximum packet latency, are not specified, since the proposed system is intended for non-time-critical applications.

Each node is placed at a maximum distance of 10 m from a neighboring node and must be able to receive and transmit a small number of packets (less than 10) in a relatively long time span (10 min). The payload of each packet is of the order of 100 bits. A Bit Error Rate (BER) of 10^{-3} is then low enough to give a packet error rate lower than 10% (for a payload of 100 bits); this is an acceptable level of performance for the applications described, which are characterized by a very low average packet rate. As explained in the next section, power dissipation in such a WSN is not dominated by data transmission and reception. High packet error rates can then be tolerated, because packet retransmissions do not significantly affect the power consumption.

2.2.2 Cost and Size

A substantial fraction of the cost of a WSN node is determined by PCB assembly and IC packaging costs. They can be significantly cut if the number of off-chip external components is reduced, or, extremely, if a fully integrated solution is adopted. An added benefit is that the minimization of the off-chip components also shrinks the size of the node.

Crystals resonators are one of the most common external components in electronic systems. Although the volume of packaged crystals may be only about 1 mm^3 [15], their size is still significant compared to the dimensions of the silicon die. Efforts have been spent in adopting alternative frequency-defining elements, such as MEMS for low-frequency real-time clocks [16, 17], or SAW [18] and BAW [19] resonators for defining the operating frequency of RF front-ends. However, these solutions can hardly be integrated or require additional cost-increasing processing steps.

For these reasons, we require the frequency reference of a practical node to be fully integrated in a standard microelectronic technology with the rest of the electronics. Among the different processes, CMOS is the best suited, thanks to the possibility of integrating complex SoCs. However, compared to crystal-based oscillators, this results in a less accurate reference. The implications of this choice will be analyzed in detail in the following.

2.2.3 Energy Consumption

To power up a WSN, energy could be distributed to each node from a nearby source. This can be accomplished, for instance, by means of light, sound waves or electromagnetic radiation. However, wireless energy transmission is often very inefficient. For example, the power that can be distributed via RF waves is constrained by both theoretical limits, such as the size of the node antenna and the distance between the nodes and the energy source, and practical limits, such as the spectrum emission regulations. In a practical situation, these limitations would result in node power levels of the order of few μW [20]. Energy distribution in this manner is then not a viable solution in most cases, though it is applicable in particular applications [21, 22].

Energy-autonomous nodes represent another solution. They can be realized either through the use of an on-board energy storage element or by "scavenging" energy from renewable energy sources in the environment. In the first case, each node is provided with a battery and its lifetime is limited by the battery capacity, since battery replacement is unpractical and expensive in ubiquitously deployed networks. In the latter, energy can be acquired from a multitude of different sources, such as solar or artificial light, mechanical vibrations and heat [23–25].

Table 2.2 reports the typical performance of batteries and energy scavengers, taking into account a lifetime of 1 year. Considering battery capacity, state-of-the-art

Table 2.2 Comparison of various power sources for WSN; data from [23]

	Energy source	Power density	Energy density
Energy scavengers	Solar	$15\,\text{mW/cm}^2$ (outdoor) $10\,\mu\text{W/cm}^2$ (indoor)	
	Vibrations	$200\,\mu\text{W/cm}^{3\text{a}}$	
	Heat	$40\,\mu\text{W/cm}^{2\text{b}}$	
Batteries[c]	Zinc-air	$120\,\mu\text{W/cm}^3$	$3.8\,\text{kJ/cm}^3$
	Lithium	$92\,\mu\text{W/cm}^3$	$2.9\,\text{kJ/cm}^3$
	Lithium (rechargeable)	$35\,\mu\text{W/cm}^3$	$1.1\,\text{kJ/cm}^3$

[a]For an acceleration of $2.25\,\text{m/s}^2$ at $120\,\text{Hz}$
[b]From two surfaces at temperature difference of 5 K
[c]Power density computed for 1 year lifetime

energy scavenging techniques, and even extrapolated future developments, the average power consumption of each node must be limited to about $100\,\mu\text{W}$.

The wireless transceiver is the most challenging subsystem to implement because its power requirements are much higher than those of other subsystems, such as the sensors and the digital processing. Thus, in the following we will consider $100\,\mu\text{W}$ as the power budget for the transceiver only. Even if the average power required by each node is lower than this, the power management system should be still able to provide a peak power greater than $100\,\mu\text{W}$. As will be shown in Sect. 2.3.1, most sub-systems are "off" most of the time, but draw significantly more than $100\,\mu\text{W}$ when turned "on." For this reasons, nodes relying on scavenged energy must also be provided with energy storage elements to continuously save incoming power and release it when needed.

2.3   The Network Synchronization

2.3.1   Asynchronous vs. Synchronous Networks

In order to reduce power consumption, a unique characteristic of WSN can be exploited, i.e. the small amount of transmitted data. With reference to the specification of Table 2.1, a packet could be transmitted using a very low effective data rate of 1.7 bps. Lowering the data rate would simplify some sections of the radio and reduce their power consumption. However, a low data rate implies a long transmission time and, hence, an increased energy consumed by those radio components whose power requirements are weakly dependent on the data rate (like the RF oscillators). It is convenient then to use a higher data rate and turn on the radio only when data packets are transmitted or received [26]. Otherwise, the largest fraction of energy consumption would be spent in idle listening to the channel, waiting for data packets [27].

Reduction of the energy wasted in idle listening is usually obtained by duty-cycling the network nodes, i.e. by putting them into a sleep mode for a significant fraction of the time. This task requires a synchronization algorithm to ensure that all nodes observe simultaneous sleep and wake-up times. Previous solutions have focused on the use of either an asynchronous [28] or a synchronous networks approach [29]. In the latter, nodes are equipped with high-accuracy time references that allows them to synchronize themselves in time. Each node can agree to use specific communication timeslots and can then be duty-cycled in order to listen to the channel only during these timeslots, thereby reducing idle listening time. However, this solution requires an accurate time reference, which usually makes use of off-chip crystal resonators.

In the reactive radio approach, nodes, in addition to a main radio for data communication, are equipped with a low-power wake-up radio. A wake-up radio is a receiver with the task of continuously listening to the channel, detecting arrival of packets and turning on the main radio for the reception of those packets. Since the wake-up radio does not need to demodulate the data, its architecture can be simplified and its power consumption can be lower than that of the main radio. Nevertheless, since it is operating continuously, the peak power consumption of the wake-up radio must be lower than $100\,\mu W$. This can be achieved with very simple architectures and by reducing the wake-up radio's sensitivity [19, 30].

In the next section, a different solution will be described, based on a *duty-cycled wake-up radio* [31]. We will show that, by using a synchronous network and a wake-up radio at the same time, it is possible to relax constraints both on the time reference accuracy and on the wake-up radio's power consumption. This allows the adoption of a fully integrated time reference and of a wake-up radio with acceptable sensitivity.

2.3.2 Duty-Cycled Wake-Up Radio

In the proposed scheme, the receiver, while residing in a reduced power mode (*sleep mode*), is able to decide when to turn itself on to listen for communications (*listening mode*), and when communications are present, to prompt a full power-up of the device (*communication mode*). This can be implemented by the architecture in Fig. 2.1a, comprising a wake-up radio responsible for monitoring the channel while waiting for data packets, a main radio to communicate, and a time reference for synchronization. The wake-up radio is duty-cycled, i.e. put in listening mode on a scheduled basis by the time reference to save energy when monitoring is not required. When a data packet is present, the wake-up radio triggers a wake-up call for the main radio. As shown in Fig. 2.1b, time is divided into fixed slots, e.g. Slot 1, Slot 2, ..., which form the basis of a Time Division Multiple Access (TDMA) Medium Access Control (MAC) protocol. Accordingly, in each timeslot only the wake-up radio of a particular node should be monitoring the channel. For example, Slot 2 can be allocated to Node 2, Slot 3 to Node 3 and so on. Any node can

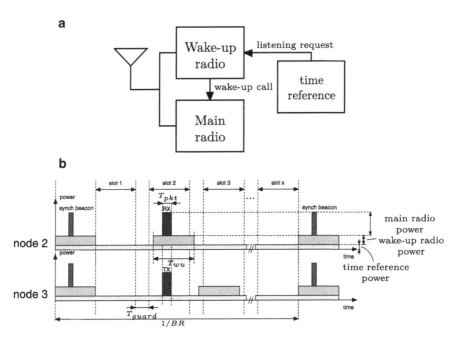

Fig. 2.1 (**a**) System architecture and (**b**) MAC protocol with energy breakdown

transmit a packet in any slot, depending on the intended recipient node. A back-off algorithm handles collisions due to the simultaneous transmission of packets from different sources to the same destination. In the figure, Node 3 sends a packet to Node 2 in Slot 2. With this approach, the wake-up radio is active when a packet is expected, while the main radio is active only when a packet is actually going to be received. Specialized timeslots are needed for the time synchronization of the whole network. In these dedicated timeslots, a particular node, called the master, sends special packets, labelled in Fig. 2.1b as "synch beacons." All nodes wake up to listen to the synchronization beacons and reset their internal time reference on reception of the beacon. In the following, we assume that a synchronization beacon has the same structure of a normal packet.

If more reliability is required, the receiver can send an "acknowledgment" when a data packet is successfully received. The acknowledgment for a packet sent by a transmitting node T to a receiving node R can be either sent as a stand-alone data packet in the next available timeslot for T or embedded in the payload of the next data packet from R to T. These approaches increase the latency in the transmission of information, but this can be tolerated, since the applications of interest in this work do not have stringent requirements on QoS. In any case, the use of acknowledgments affects the power consumption only marginally, since, as it will be shown in the following sections, the power is not dominated by data packet transmission and reception.

In the next section, it will be shown that the power consumption needed for an implementation of a TDMA scheme depends on the time reference accuracy. The adoption of a TDMA scheme in combination with a duty-cycled wake-up radio reduces the power consumption for a given time reference accuracy and consequently makes it possible to use a fully-integrated time reference with moderate inaccuracy, i.e. 5,000 ppm or less. Moreover, the sensitivity of the wake-up radio can be increased, since its power budget is increased by duty-cycling.

2.3.3 Optimization of MAC Performance

The average power consumption of each node depends both on the power consumption of each block (main radio, wake-up radio and time reference) and their duty-cycle factors. To determine the latter, it is necessary to determine the parameters of the MAC protocol, such as the duration of a timeslot and the rate of synchronization beacons, and then optimize the duty-cycle factors for the minimum power consumption.

The duration of the listening timeslots (T_{wu}) must be long enough to account for timing errors between the time references of the receiver and the transmitter, thus ensuring that a packet is only transmitted when the recipient is in listening mode. Since the time references are reset by the synchronization beacons, timing errors in receivers and transmitters are accumulated from the last beacon and depend on time reference accuracy. We assume that the timing error of each time reference is bounded by the interval $[-a_{time}t, +a_{time}t]$, where a_{time} is the relative time reference inaccuracy and t is the time that has elapsed since the last synchronization. Taking into consideration the fact that T_{wu} must comprise the time needed to receive at least n_{pkt} packets and a margin to account for the timing error, the following condition must hold:

$$T_{wu} \geq \frac{4a_{time}}{BR} + n_{pkt}T_{pkt} \qquad (2.1),$$

where BR is the beaconing rate, i.e. the repetition rate of the synchronization beacons, T_{pkt} is the packet transmission time and n_{pkt} is the number of packets that can be received in one slot.[3] The factor 4 in (2.1) can be understood by considering the two worst cases: if the transmitter is early (late) by a_{time}/BR and the receiver is late (early) by the same amount, the wake-up radio must be on at $2a_{time}/BR$ before (after) the time expected according to the receiver time reference.

---

[3]Equation (2.1) gives a sufficient condition for enabling communication among all nodes. However, nodes whose timeslots occur shortly after a synchronization beacon can, in principle, have a shorter T_{wu} and consequently a lower power consumption. This case is not analyzed in this work. In the following it is implicitly assumed that T_{wu} is the same for all nodes.

However, T_{wu} can not be arbitrarily large, as the protocol must assure the availability of a timeslot for each node. This is expressed by the following condition:

$$\frac{1}{PR} \geq \left(n_{nodes} + \frac{BR}{PR}\right)\left(T_{wu} + T_{guard}\right) \tag{2.2}$$

where PR is the packet rate of each node, n_{nodes} is the number of nodes in the network and T_{guard} is the guard time between two successive timeslots. Equation (2.2) ensures that the interval $1/PR$ is wide enough to accommodate timeslots for reception of data packets and synchronization beacons.

The packet transmission time can be expressed as $T_{pkt} = \frac{N_{pr}+N_{pl}}{DR}$, where N_{pr} and N_{pl} are, respectively, the number of bits in the preamble and in the payload of each packet, and DR is the data rate of the main radio. A high data rate would reduce the packet transmission time and consequently the duty-cycle of the main radio; however, power-hungry transceivers would be required to achieve very high data rates. On the other hand, low data rates and long transmission times can also be inefficient because in these cases the power in the transceiver would be dominated by data-rate independent blocks, such as RF oscillators. As a good trade-off, $DR = 100$ kbps is chosen [32].

The duty-cycle of the wake-up radio and of the transmitting and receiving sections of the main radio can be expressed, respectively, as:

$$DC_{wu} = T_{wu} \left(PR + BR\right) \tag{2.3}$$

$$DC_{rx} = \frac{N_{pr} + N_{pl}}{DR} \left(PR + BR\right) \tag{2.4}$$

$$DC_{tx} = \frac{N_{pr} + N_{pl}}{DR} PR \tag{2.5}$$

The average node power consumption is then defined as

$$P = DC_{wu}P_{wu} + DC_{rx}P_{rx} + DC_{tx}P_{tx} + DC_{clk}P_{clk} \tag{2.6}$$

where P_x is the peak power consumption of block x and it is assumed that $DC_{clk} = 100\%$ for the duty-cycle of the time reference.

Using the previous equations, it is possible to find the optimum BR which minimizes the power consumption and satisfies conditions (2.1) and (2.2). The results are shown in Fig. 2.2, in which the optimum BR and the minimum power consumption for the radio section (wake-up radio and main radio) are plotted as a function of the time reference inaccuracy. We have assumed $P_{wu} = 500\,\mu\text{W}$, $P_{rx} = P_{tx} = 2\,\text{mW}$, $PR = 1$ pkt/min $N_{pr} = N_{pl} = 100$, $T_{guard} = 0.25T_{wu}$ and $n_{pkt} = 1$ in (2.1). These assumptions will be justified in the following sections. The power dissipation of the time reference is not included in this analysis, as it was not possible to find a simple relation between its accuracy and its power consumption. Further considerations on a practical implementation are given in Sect. 2.3.5.

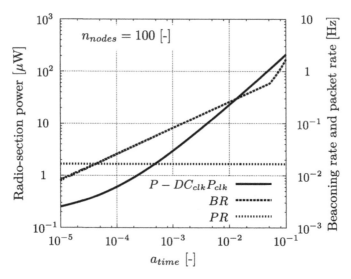

Fig. 2.2 Minimum average power consumption of the radio section (wake-up radio and main radio) and relative beacon rate vs. time reference inaccuracy for a network of 100 nodes; for reference, packet rate is also plotted

Figure 2.2 shows that the power increases as the time reference inaccuracy increases. This can be expected as large a_{time} imply either longer timeslots or more frequent beacons. BR is then optimized for the best balance between power spent in the wake-up radio, in the case of longer timeslots, and the power spent in the whole radio section in the case of more beacons. Network synchronization plays a crucial role in the estimation of the total power budget and in most cases it also represents the dominant fraction of the power for this simple MAC scheme. In Fig. 2.2, BR is higher than PR for a_{time} higher than 40 ppm, which means that nodes are receiving beacons more often than data packets. It would then be more efficient to have a highly accurate time reference; however, the analysis in this section shows that a power consumption compatible with system specifications is achieved also with relaxed accuracy and, consequently, with a fully integrated solution.

A discontinuity of the derivative of BR is observable in Fig. 2.2. This can be explained by the fact that for high values of a_{time}, very frequent beacons are needed to keep T_{wu} low enough to assign a timeslot to each node [see (2.1)]. The minimum power for the radio section and the network activity are plotted vs. the number of nodes in Fig. 2.3. The activity of the network is defined as the ratio of the time during which at least one node is listening or receiving (comprising also the guard time) and the total available time. As n_{nodes} increases, the activity saturates to 100%. The average power also rises until a limit in the number of nodes is reached for which is not possible to satisfy conditions (2.1) and (2.2). This limit in the maximum number of nodes is clearly dependent on the accuracy of the time reference and represents a limit to the scalability of the network. Moreover, the computations have been performed with a simple model where no errors, such as packet collisions,

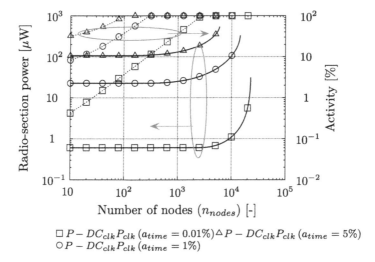

Fig. 2.3 Average power consumption of the radio section (wake-up radio and main radio) and network activity vs. number of nodes; *curves* for three different values of time reference inaccuracy are reported

have been considered. We can take collisions into account by reserving time for the reception of more packets in each timeslot: for $a_{time} = 1\%$ we obtain a maximum n_{nodes} of about 14.400 and 3.600 nodes respectively for $n_{pkt} = 1$ and $n_{pkt} = 4$. Other sources of errors must be considered to give a more realistic prediction for the power P; in the next section, the loss in power introduced by non-idealities of the wake-up radio is calculated.

2.3.4 Impact of Wake-Up Radio Non-idealities

The operation of the wake-up radio is outlined here: the wake-up radio listens to the channel for a fixed time T_d and decides if a useful signal is being received or not; if a wake-up request is detected, the main radio is turned on and it tries to synchronize to the incoming signal for the duration of the preamble[4]; if synchronization succeeds, the payload is processed; otherwise the main radio is turned off, control is handed back to the wake-up radio and the whole operating cycle is repeated. We define a *false alarm* as the waking-up of the main radio when no useful signal has been received and a *missed detection* as a failure to wake up when a packet should have been received. If a node does not wake up to receive a packet, it will be retransmitted in a successive timeslot. The performance of the wake-up radio is

---

[4]For more details on synchronization, see Sect. 2.4.3.

then completely specified in terms of the probability that a particular decision will result in a false alarm ($\mathbb{P}_{fa}$), the probability that a particular decision will result in a missed detection ($\mathbb{P}_{md}$) and the time required to make these decisions T_d.

Taking these non-idealities into account, (2.4) and (2.5) have to be modified as:

$$DC_{rx} = \frac{N_{pr} + N_{pl} + n_{fa}N_{pr}}{DR}(PR + BR) \tag{2.7}$$

$$DC_{tx} = (n_{md} + 1)\left(T_d + \frac{N_{pr} + N_{pl}}{DR}\right)PR \tag{2.8}$$

where n_{fa} is the average number of false alarms issued per timeslot and n_{md} is the average number of missed detections per packet. Note that the transmitter must send a wake-up request in an additional time T_d before each packet in order to wake-up the receiver.

The number of false alarms can be approximated as[5]:

$$n_{fa} \approx \frac{\mathbb{P}_{fa}T_{wu}}{T_d + \dfrac{N_{pr}\mathbb{P}_{fa}}{DR}} \tag{2.9}$$

while we assume that the number of missed detection is

$$n_{md} = \mathbb{P}_{md} \tag{2.10}$$

This approximated expression is in good accordance with simulations, as shown in Fig. 2.4 where (2.9) and the results of simulations are plotted for the case $T_{wu} = 200$ ms and $DR = 100$ kbps.

In order to find the requirements for $\mathbb{P}_{fa}$, $\mathbb{P}_{md}$ and T_d, we should know the relation between the power dissipated in the wake-up radio P_{wu} and the probability of errors. The knowledge of such relation would enable the full optimization of total power P: a lower probability of errors would decrease the duty-cycle of the main radio but the wake-up radio would then need to burn more power to achieve this. Moreover, a trade-off exists for the choice of T_d: a longer T_d increases the transmission time but, simultaneously, reduces the probability of wake-up radio errors, as will be shown in Sect. 2.5.2. Taking into account a possible implementation of the wake-up radio (see Sect. 2.5.2), the following parameters have been chosen: $T_d = 5.24\,\mu s$, $\mathbb{P}_{fa} = 2.6 \cdot 10^{-5}$ and $\mathbb{P}_{md} = 10^{-2}$, resulting in $n_{fa} \approx 1$ and $n_{md} = 10^{-2}$.

In the previous analysis, the case in which a missed detection occurs during the reception of a synchronization beacon has been neglected. The node would simply

---

[5]Equation (2.9) is not mathematically rigorous; it has been developed considering that n_{fa} is the product of the number of decisions to be taken in a timeslot by $\mathbb{P}_{fa}$ and that the number of decisions depends on the difference between the duration of the timeslots and the time used for reception of the preamble caused by false alarms, i.e. $n_{fa} = \mathbb{P}_{fa}\frac{T_{wu} - n_{fa}N_{pr}/DR}{T_d}$.

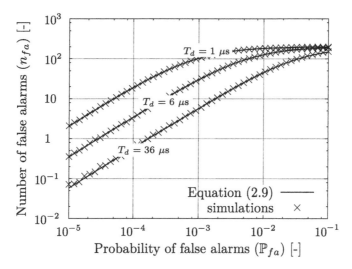

Fig. 2.4 Average number of false alarms per timeslots for different values of T_d for the case $T_{wu} = 200\,\text{ms}$ and $DR = 100\,\text{kbps}$

Table 2.3 MAC parameters

Parameter	Value
DR	100 kbps
BR	0.2 Hz
T_{wu}	200 ms
T_{guard}	50 ms
N_{pr}	100
N_{pl}	100
$\mathbb{P}_{fa}$	$2.6 \cdot 10^{-5}$
$\mathbb{P}_{md}$	10^{-2}
a_{time}	1%
n_{pkt}	4

skip that particular beacon and increase its T_{wu} until the reception of the next one. The effect on the power consumption is negligible and has been neglected in the above expressions.

2.3.5 System Performance

The optimum parameters for the MAC scheme have been chosen using the guidelines described previously and are reported in Table 2.3. To carry out the optimization, estimates have been made of the peak power dissipated in the different subsystems, as was already briefly mentioned in Sect. 2.3.3, and the accuracy of the time reference has been fixed.

Table 2.4 System performance for time reference inaccuracy of 1%

Subsystem	Peak power	Duty-cycle (%)	Average power
Main radio (Rx)	2 mW	0.065	1.3 μW
Main radio (Tx)	2 mW	0.003	0.1 μW
Wake-up radio	500 μW	4.333	21.7 μW
Time reference	50 μW	100	50.0 μW
Total system			73.0 μW

The power requirements of the three sub-blocks (main radio, wake-up radio and time reference) are discussed in Sect. 2.5 and are reported in Table 2.4. The inaccuracy of the time reference is fixed to 1% and the reasoning behind this limit is shortly presented in Sect. 2.5.3 and broadly analyzed in the next chapters.

With these assumptions for the peak power and with the duty-cycle factors derived from the MAC analysis for a time reference inaccuracy of 1%, the average power consumption can be computed and is reported in Table 2.4. The total average power is smaller than the allowed budget of 100 μW, demonstrating the feasibility of the system in WSN applications.

This analysis of the duty-cycled wake-up radio MAC protocol has shown the effect of time reference inaccuracy on the power consumption of WSN node. It has also been proven that the overhead due to network synchronization can become the dominant fraction of power consumption when a fully integrated time reference is adopted. Despite the low complexity of the proposed protocol, similar issues would arise when low-accuracy time references are used in more sophisticated protocols, such as T-MAC [33] or WiseMAC [29].

2.3.6 Master Node

A final remark should be made about the power consumption of the master node. The master node is equipped with the same hardware as the other nodes in the network: it is one of the nodes in the network which is elected as master during the setup phase of the network (not described in this work). Its power consumption differs slightly from that shown in the power budget of Table 2.4, since it does not receive synchronization beacons, but transmits them. Since we considered equal power for transmitting and receiving, the only difference in power consumption between the master node and any other node, is the power spent in the wake-up radio while waiting for the beacon. This corresponds to a saving of 20 μW and a total average power for the master node of 53 μW.

2.4 Frequency Accuracy

Frequency accuracy is needed in the main radio to enable RF communication. According to the adopted modulation scheme, frequency accuracy is needed both at the transmitter side to ensure transmission in the right portion of the spectrum and at the receiver side to be able to find the frequency at which the incoming signal is placed. In the following sections the choice of the modulation scheme most suited in case of a low frequency accuracy and the accuracy requirements for the transmitter and receiver are derived.

2.4.1 RF Modulation

The RF modulation scheme must be chosen to relax the frequency accuracy requirements enough to enable the use of a low-accuracy fully integrated frequency reference. In a receiver, the allowed frequency error is directly proportional to the bandwidth of the RF signal. For narrowband modulation schemes (such as OOK, FSK or QAM), the specifications on frequency accuracy are then very strict, since the signal bandwidth is of the same order of the low data rate adopted in WSN. More suitable is an Ultra Wide Band (UWB) system employing a modulation scheme based on Impulse Radio (IR) [34]; the RF signal occupies a bandwidth of hundreds of MHz to comply with radio regulations, and a fully integrated reference for this application can be easily built. However, it will be difficult to meet the power requirements at the receiver due to the inherent wideband nature of an UWB receiver. A better solution is the use of an Impulse Radio signal with a bandwidth smaller than that required in UWB systems but large enough to relax frequency accuracy constraints for full integration [31].

The preferred frequency band is the 2.4-GHz ISM-band, which allows occupation of tens of MHz at frequencies high enough to enable the integration of the required passive components (e.g. inductors) on chip. Figure 2.5 shows the simplified time representation of the adopted signal and its spectrum. An RF carrier is modulated by a pulse waveform with period T_f and duty cycle T_p/T_f. The pulses are shaped as square waves with duration T_p and pulse repetition frequency $PRF = 1/T_f$. Each bit is represented by a sequence of $n_p = PRF/DR$ successive pulses, where $DR = 1/T_b$ is the data rate, implementing a simple repetition code.[6] The bits are modulated using Pulse Position Modulation (PPM). In each frame, i.e. in each slot of duration T_f the pulse can be transmitted with different delays: the pulse can be positioned with zero delay (bit 0) or with delay T_{ppm} (bit 1), in case of

---

[6]The repetition code is a coding scheme in which each data bit is transmitted multiple times over the channel.

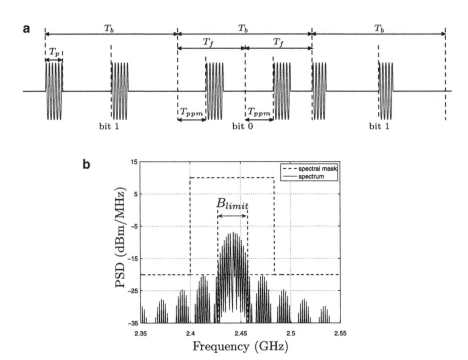

Fig. 2.5 Example of IR signal (**a**); spectrum with parameters of Table 2.5 (**b**)

binary modulation; more delays can be added to employ an M-ary modulation. In the following, constraints affecting the modulation parameters choice are listed in order to find an optimal set of parameters.

2.4.2 Transmitter Limits

With an average transmitted power[7] $P_{avg} = 1$ mW, the European ETSI 2.4-GHz ISM band requirements are met if the transmission frequency resides in the interval $[f_0 - \Delta f_{TX}, f_0 + \Delta f_{TX}]$, where $f_0 = 2.44175$ GHz is the nominal transmitting frequency at the center of ISM band and

$$\Delta f_{TX} = 41.75\,\text{MHz} - \frac{B_{limit}(T_p)}{2} \tag{2.11}$$

where the numerically computed B_{limit} is the width of the spectrum of Fig. 2.5b at -20 dBm/MHz.

---

[7] The power of the transmitted signal is limited by the peak power provided by the energy source and the power management system and the expected efficiency of the transmitter. Note that the spectral mask requirements are still met if less power than 1 mW is emitted and the assumed value is a practical upper bound.

The complexity of transmitter circuitry depends on the signal Crest Factor, defined as $CF \triangleq \frac{P_{peak}}{P_{avg}} = \frac{T_f}{T_p}$ where P_{peak} is the peak power.[8] In order not to put excessive requirements on the transmitter, it has been chosen to limit the crest factor to less than $CF_{TX,max} = 10$. The following condition must hold:

$$CF = \frac{T_f}{T_p} \leq \min \left\{ CF_{TX,max}, \frac{P_{peak,limit}}{P_{avg}} \right\} = 10 \qquad (2.12)$$

where $P_{peak,limit}$ is fixed by the regulations.

The signal power at the receiver antenna can be computed using the macroscopic model for the path loss [36]. The path loss is given by

$$PL(d) = \left(\frac{4\pi}{c} r_0 f \right)^2 \left(\frac{d}{r_0} \right)^{n_{PL}} \qquad (2.13)$$

where c is the speed of light, $r_0 = 1$ m is a reference distance, d is the distance between transmitter and receiver antenna, f is the signal frequency and n_{PL} is the path loss coefficient. For indoor propagation, n_{PL} is usually in the range between 3 and 4. In the worst case ($n_{PL} = 4$), with the parameters previously defined ($d = 10$ m, $f = 2.44$ GHz) the path loss is 80 dB. Taking into account an additional margin of 4 dB, this corresponds to a line-of-sight path loss ($n_{PL} = 2$) of 60 dB and to a margin of 24 dB for fading and obstructions effect. Since the transmitted power is $P_{avg} = 1$ mW, the specification on receiver sensitivity is -84 dBm.

2.4.3   Receiver Limits

To simplify the architecture, we assume the use of repetition coding and of a non-coherent receiver. In the following analysis, the receiver is modeled as a matched filter followed by an ideal sampler [37]. With the adopted PPM scheme, the output of the matched filter is sampled twice per frame, i.e. in the position for bit 0 and for bit 1, and the decision on the received bit is performed in the digital domain using a majority criterion: the bit is chosen according to the majority of the values of the demodulated pulses. For a fixed P_{avg} and DR, the implementation loss related to this simple decoding algorithm increases with the number of pulses per bit n_p; consequently the following practical limit is posed:

$$n_p = \frac{T_b}{T_f} \leq 25 \qquad (2.14)$$

---

[8]Transmitting a signal with higher peak power requires a transmitting chain and, in particular, a power amplifier with higher linearity. Since linearity and power efficiency of a transmitter are contrasting requirements [35], a modulation requiring a lower P_{peak} and consequently a lower CF is preferred.

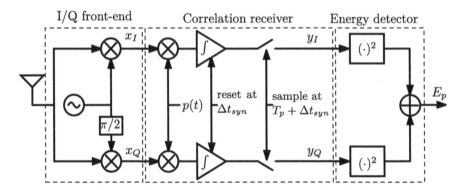

Fig. 2.6 Block diagram of the receiver

which corresponds to an implementation loss due to coding of 5.1 dB in receiver sensitivity for a Bit Error Rate (BER) of 10^{-3}. The loss can be overcome using a more complex coding scheme than repetition coding. If no coding gain is required but only the recovery of the loss, very simple schemes could be employed, which do not require complex digital hardware, but that is out of the scope of this work. Note that in terms of BER performance PPM is equivalent to the classical On-Off Keying (OOK) modulation.

The receiver synchronizes itself in time and frequency to the incoming signal using the N_{pr} bits of the preamble.[9] In case an imperfect synchronization is achieved, an implementation loss must be taken into account. It can be computed considering the equivalence between a matched filter and a correlation receiver and the need for two quadrature branches in a non-coherent receiver to tackle the phase difference between incoming signal at the antenna and local oscillator (Fig. 2.6). Denoting with x_I and x_Q and with y_I and y_Q the in-phase and quadrature components respectively of the input and the output of the in-phase and quadrature correlation receivers, the detected energy of a single pulse after sampling can be expressed as[10]

$$E_p \left(\Delta f_{syn}, \Delta t_{syn} \right) =$$
$$= y_I^2 + y_Q^2$$

---

[9]The number of bits in the preamble is chosen large enough to ease the implementation of the synchronization algorithm. At the same time, the choice $N_{pr} = 100$ does not affect sensibly the total power consumption (see Sect. 2.3.5).

[10]Note that in the following we compute the detected energy of only one sampling, but in case of PPM two samplings per frame are needed. The adopted simplification is possible thanks to the equivalence in terms of energy and BER performance of PPM and OOK.

$$= \left(\int_{\Delta t_{syn}}^{\Delta t_{syn}+T_p} x_I(t) p(t) dt \right)^2 + \left(\int_{\Delta t_{syn}}^{\Delta t_{syn}+T_p} x_Q(t) p(t) dt \right)^2$$

$$= \frac{1 - \cos\left[2\pi \Delta f_{syn} \left(T_p - |\Delta t_{syn}| \right) \right]}{2 \left(\pi \Delta f_{syn} \right)^2} \tag{2.15}$$

where Δf_{syn} and Δt_{syn} are respectively the error in frequency and timing between the actual value and the estimated one at the receiver, $p(t)$ is the pulse shape, defined for our modulation scheme as

$$p(t) = \begin{cases} 1, & \text{if } t \in [0, T_p] \\ 0, & \text{otherwise} \end{cases} \tag{2.16}$$

x_I and x_Q are given by

$$\begin{cases} x_I = p(t) \cos(2\pi \Delta f_{syn} t + \phi) \\ x_Q = p(t) \sin(2\pi \Delta f_{syn} t + \phi) \end{cases} \tag{2.17}$$

where ϕ is the constant phase difference between received signal and local oscillator. Since the performance of the matched filter receiver depends on the received energy, the implementation loss is

$$IL_{syn} = \frac{E_p(0,0)}{E_p \left(\Delta f_{syn}, \Delta t_{syn} \right)}$$

$$= \frac{2 \left(\pi \Delta f_{syn} T_p \right)^2}{1 - \cos\left[2\pi \Delta f_{syn} \left(T_p - |\Delta t_{syn}| \right) \right]} \tag{2.18}$$

2.4.4 Optimization

The choice of T_p depends on Δf_{osc}, defined as the error of the local oscillator frequency with respect to the nominal frequency. All previous constraints are plotted in Fig. 2.7:

- The transmitter limit (solid line), i.e., with reference to (2.11),

$$\Delta f_{osc} \le \Delta f_{TX} \tag{2.19}$$

- The CF limit (dashed-dotted line), i.e.

$$\frac{T_f}{T_p} \cdot \frac{T_b}{T_f} \le 10 \cdot 25 \Rightarrow T_p \ge \frac{T_b}{250} = 40 \, \text{ns} \tag{2.20}$$

derived using (2.12) and (2.14), where T_b is assumed to be $10 \, \mu s$ ($DR = 100 \, \text{kbps}$) in our application;

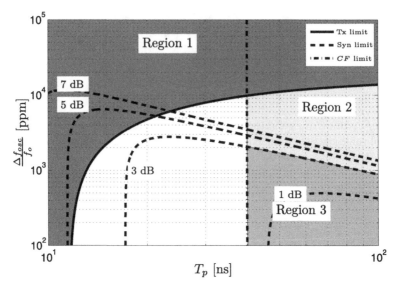

Fig. 2.7 Plot for the choice of T_p; different constraints are shown: transmission limit (2.19), CF limit (2.20) and synchronization limit (2.21) for various IL_{syn} (shown on the *curves*)

- The synchronization limit (dashed line), i.e., with reference to (2.18),[11]

$$\Delta f_{osc} \leq \frac{\Delta f_{syn}}{2} \tag{2.21}$$

plotted for different values of IL_{syn} and a fixed timing error for the synchronization algorithm ($|\Delta t_{syn}| = 5$ ns).

If the accuracy of the oscillator is enough for a given IL_{syn}, no frequency estimation needs to be performed at the receiver. Since it is possible to trade off timing error (Δt_{syn}) for frequency error (Δf_{syn}) in (2.18), a small timing error (5 ns) has been chosen. This is advantageous in terms of hardware as it is more difficult to tune the frequency of the receiver than the timing.

From Fig. 2.7 a good choice for T_p can be found using frequency accuracy considerations. Referring to an IL_{syn} of 3 dB, it is possible to distinguish different regions. In region 1, i.e. all the points above the transmission limit, the transmitter will not respect the spectral mask. Points in region 2 respect the transmission limit but some frequency synchronization at the receiver is needed because the frequency accuracy is not good enough. Region 3, i.e. the points under the transmission limit and under the curve of $IL_{syn} = 3$ dB, contains points for which no frequency synchronization is required at the receiver because the frequency accuracy provided by

---

[11]The factor 2 derives from the presence of frequency errors both in transmitter and receiver.

Table 2.5 Modulation
parameters

Parameter		Value
DR		100 kbps
P_{avg}		1 mW
−3 dB bandwidth		17.7 MHz
CF		9.52
IR parameters	T_f	476 ns
	T_{ppm}	238 ns
	T_p	50 ns
Synchronization	IL_{syn}	3 dB
parameters	Δf_{syn}	8.4 MHz
	Δt_{syn}	5 ns

the local oscillator is enough to maintain IL_{syn} below 3 dB. The optimal point is strictly related to the available oscillator frequency accuracy. The optimal point in terms of relaxed frequency accuracy is obtained for $T_p = 40$ ns, in which case an inaccuracy of 0.2% is required to keep IL_{syn} below 3 dB. However, if an accurate frequency reference, e.g. a crystal oscillator, is not available, it will be challenging to achieve accuracies below 1%, as it will be shown in the following chapters. At the 1% level of inaccuracy, a frequency synchronization algorithm must be used for any value of T_p, in order to keep the implementation loss due to the synchronization system less than 3 dB. A choice of $T_p = 50$ ns has been made to keep some margin from the CF limit, leading to an allowed absolute timing and frequency error equal to 5 ns and 8.44 MHz respectively.

The frame period T_f is chosen accordingly to different requirements: (2.12), (2.14) and $PRF < 2.5$ Mpps (to avoid a very fast baseband). Thus, $T_f = 476$ ns is employed and consequently, $PRF = 2.1$ Mpps, $n_p = 21$ and $CF = 9.52$. The resulting modulation parameters are reported in Table 2.5.

With the parameters adopted for the IR modulation and 1% frequency inaccuracy, each transmitting node respect the spectral regulations and a small CF factor is specified for the implementation of the transmitter. At the receiver side, a synchronization algorithm must be adopted to tune the receiver frequency and achieve a frequency error of the order of 0.1% ($IL_{syn} < 3$ dB). Since the required accuracy for the synchronization algorithm is one tenth of the initial error, also a very simple linear search in the frequency space would require only 10 steps.

2.5 Node Architecture

After discussing the impact of the frequency reference accuracy on both the network synchronization and the RF transmission, the feasibility of the different sub-blocks of the WSN node must be proven. A short overview of the features and implementation issues of the main radio, the wake-up radio and the time reference is given in the following.

Table 2.6 Main radio
requirements

Parameter	Value
Power consumption—TX (P_{tx})	2 mW
Power consumption—RX (P_{rx})	2 mW
Sensitivity (S)	−84 dBm
Bit error rate (BER)	10^{-3}
Data rate (DR)	100 kbps
Operating frequency	2.4 GHz

2.5.1 Main Radio

The specifications for the main radio have been derived in the previous sections and
are summarized in Table 2.6. The power consumption of the main radio in both
receive and transmit modes is estimated to be in the order of a few mW, taking
into consideration the state-of-the-art [38], the required sensitivity and a transmitted
power of 1 mW (see Sect. 2.4.2). However, basic parameters for transceiver design,
such as interferers immunity or synchronization capabilities, have been neglected
in this analysis and are out of the scope of this work. Though such additional
constraints can be of secondary relevance for the low-data-rate low-performance
radio link typical of WSNs, considering them among the radio specifications can
bring to an increase of the complexity of the RF front-end and consequently to a
higher power consumption.

Despite that, the contribution of the main radio to the total average power is
marginal and errors in the estimation of P_{rx} and P_{tx} can be largely tolerated. Even
if the peak power for transmitter and receiver would be tenfold, their contribution to
the power budget would only represent a fraction smaller than 17%.

2.5.2 Wake-Up Radio

Architecture and Performance

Wake-up radios can have a very low power consumption thanks to their simplified
functionality with respect to the main radio. However, if the wake-up radio were
not duty-cycled, it would be challenging to meet a power specification lower than
100μW while keeping an acceptable sensitivity [19, 30]. Our solution overcomes
these limits at two different levels. At the network level, the adoption of a TDMA
scheme allows duty-cycling of the wake-up radio, relaxing in this way the requirements on its instantaneous power dissipation, which can be as high as 500μW, as
proven in Sect. 2.3. The increased power budget available for the wake-up radio will
reduce the noise added by the circuitry and increase the sensitivity. At the physical
level, the use of Impulse Radio is beneficial for the wake-up functionality. Since
the peak power of the pulses is higher than the average power of the incoming
signal, it is easier for the wake-up radio to discriminate in the amplitude domain
a useful signal from noise. Thus, on the one hand, the higher available power allows

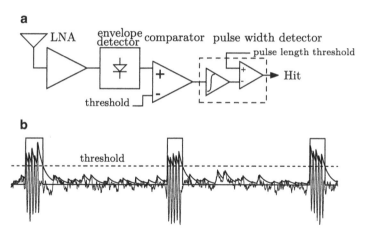

Fig. 2.8 Wake-up radio architecture (**a**) and example related waveforms (**b**): the outputs of LNA, envelope detector and comparator are shown

Table 2.7 Wake-up radio requirements

Parameter	Value
Power consumption (P_{wu})	$500\,\mu W$
Sensitivity (S)	$-84\,dBm$
Probability of false alarm ($\mathbb{P}_{fa}$)	$2.6 \cdot 10^{-5}$
Probability of missed detection ($\mathbb{P}_{md}$)	10^{-2}
Wake-up decision time (T_d)	$5.24\,\mu s$

the design of circuits with lower noise and, on the other hand, the system is more immune to the effect of noise thanks to IR modulation. We will then prove the feasibility of a wake-up radio with the specifications drawn in the previous sections.

A possible implementation of a wake-up radio and the representation of signals for each block are shown in Fig. 2.8a,b respectively. When the signal of Fig. 2.5a is present at the antenna and an appropriate threshold is chosen, the output of the envelope detector consists of a noisy train of pulses down-converted to the baseband. A comparator is used to discriminate the useful signal from the noise and in case of a clean signal its output is a square wave with the same shape of the envelope of the IR signal. A pulse width detector verifies that the pulses are longer than a certain pulse length threshold in order to prevent the system to give false alarms in case of short and high noise peaks.

As for a standard radio, it is possible to define the sensitivity of a wake-up radio. The sensitivity is defined as the minimum signal level that the system can detect with acceptable performance. The requirement on the sensitivity for the main radio has been fixed to $-84\,dBm$ in Sect. 2.4.2 and we adopt the same for the wake-up radio. Other specifications have also been defined in Sect. 2.3.4 in terms of $\mathbb{P}_{fa}$, $\mathbb{P}_{md}$ and T_d and are summarized in Table 2.7. For the architecture in Fig. 2.8a, the sensitivity is given by [39]:

$$S = kT \cdot NF \cdot B_n \cdot SNR \qquad (2.22)$$

where k is the Boltzmann constant, T is the absolute temperature, B_n is the receiver noise bandwidth, NF is the noise figure of the LNA and SNR is the minimum signal-to-noise ratio at the input of the envelope detector, which is needed to achieve the required $\mathbb{P}_{fa}$ and $\mathbb{P}_{md}$. Once the minimum SNR is known, from (2.22) it is possible to find NF, which allows to estimate the power consumption of the wake-up radio. The minimum required SNR will be computed in the following.

We define the *missed pulse probability* (p_{mp}) as the probability to miss a single incoming pulse and the *false pulse probability* (p_{fp}) as the probability to detect a noise peak as an incoming pulse. The relation between p_{fp} and p_{mp} and the network level parameters $\mathbb{P}_{fa}$ and $\mathbb{P}_{md}$ is easy to find in the case when m pulses are transmitted to wake up the receiver. We suppose that the wake-up radio of Fig. 2.8a checks the presence of a pulse inside each frame of duration T_f. The digital output of the wake-up radio (Hit) is high when a pulse is detected and after each frame the system stores the value of the digital output. After the observation of m frames periods, the number of hits, i.e. the number of times a pulse was detected in m frames, is counted. If that number is larger than a fixed value n_{th}, a wake-up call to the main radio is issued. Otherwise the count is reset and the procedure is started again. In this scenario $\mathbb{P}_{fa}$ and $\mathbb{P}_{md}$ are given by the following equations:

$$\mathbb{P}_{fa} = \sum_{k=n_{th}+1}^{m} \binom{m}{k} p_{fp}^k (1 - p_{fp})^{m-k} \tag{2.23}$$

$$\mathbb{P}_{md} = \sum_{k=0}^{n_{th}} \binom{m}{k} (1 - p_{mp})^k p_{mp}^{m-k} \tag{2.24}$$

Equations (2.23) and (2.24) are plotted in Fig. 2.9 in the case of $m = 11$, which corresponds to $T_d = 5.24\,\mu s$ (see Table 2.5).

Matlab simulations were performed to find the relation between p_{fp}, p_{mp} and SNR for the architecture of Fig. 2.8a. They can be computed by evaluating the probability of a hit in a frame where a pulse is present and the probability of a hit in a frame with no pulses but only noise. An IR waveform with a signal power equal to the sensitivity of the main radio ($S = -84\,dBm$) at 2.44 GHz and with a $CF = 9.52$ ($T_f = 476\,ns$, $T_p = 50\,ns$) was fed to the antenna. An Additive White Gaussian Noise (AWGN) channel was assumed and the SNR at the input of the envelope detector was swept.[12] The threshold of the pulse width detector is set to 45 ns, the envelope detector has an exponential discharge behavior with time constant of 10 ns and the comparator has a bandwidth of 40 MHz. The threshold of

---

[12]Though the analysis was carried on with an $AWGN$ channel, it must be noted that a margin for fading and obstruction has been considered in Sect. 2.4.2. Propagation through a multipath channel only affects the energy received for each pulse, since the delay between successive pulses is much higher than multipath delay expected by typical channel models. The effect of interferers is treated in the next section.

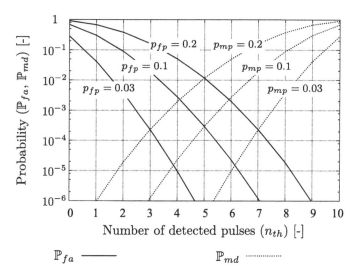

Fig. 2.9 Probability of false alarm $\mathbb{P}_{fa}$ and probability of missed detection $\mathbb{P}_{md}$ vs. number of detected pulses n_{th}; *curves* are plotted for fixed values of $m=11$ and the values of p_{fp} and p_{mp} reported in the figure

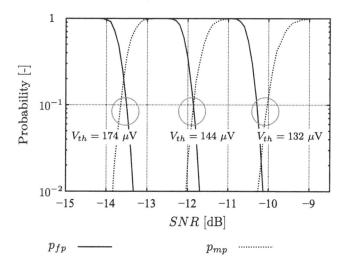

Fig. 2.10 Probability of false pulse and probability of missed pulse vs. signal to noise ratio SNR; *curves* are plotted for different values of V_{th}

the comparator V_{th} has a dramatic impact on the performance of the wake-up radio: it should be high enough to minimize p_{fp} and low enough to minimize p_{mp}. To determine the effect of V_{th} on the wake-up radio performance, p_{fp} and p_{mp} were computed with different threshold values and results are summarized in Fig. 2.10.

An optimum choice for n_{th} and V_{th} exists, which minimizes the SNR required to achieve $\mathbb{P}_{fa}$ and $\mathbb{P}_{md}$ in Table 2.7. From inspection of Fig. 2.9, it may be seen that, with $n_{th} = 6$, $p_{fp} = 10^{-1}$ and $p_{mp} = 10^{-1}$, the network requirements on $\mathbb{P}_{fa}$ and $\mathbb{P}_{md}$ are satisfied. Figure 2.10 shows that, with the choice $V_{th} = 144\,\mu V$, an SNR of $-11.8\,dB$ is required to achieve both the condition $p_{fp} < 10^{-1}$ and $p_{mp} < 10^{-1}$. It can be proven that the choice $n_{th} = 6$ and $V_{th} = 144\,\mu V$ is a good approximation of the minimum SNR. If $V_{th} > 144\,\mu V$ is chosen, there are no values of SNR and n_{th} for which both the conditions on $\mathbb{P}_{fa}$ and $\mathbb{P}_{md}$ are met. Finally, if $V_{th} < 144\,\mu V$, a range of values for n_{th} exists for which the system over-performs the requirements in terms of $\mathbb{P}_{fa}$ and $\mathbb{P}_{md}$. However, in that case a higher SNR is required, which requires more power consumption in the front-end.

The optimum SNR is dependent on the decision time T_d. A lower SNR can be obtained by increasing the number of pulses m and appropriately adjusting the threshold n_{th}: in that case, higher p_{fp} and p_{mp} can be allowed to achieve a fixed $\mathbb{P}_{fa}$ and $\mathbb{P}_{md}$ (see Fig. 2.9). On the other hand, higher p_{fp} and p_{mp} allows a lower SNR by increasing the threshold V_{th} (see Fig. 2.10) and, consequently, a higher NF for the LNA can be accepted while maintaining the same performance. Even though a higher value of m decreases the power of the wake-up radio, it results in a longer T_d, which, as was already discussed in Sect. 2.3.4, increases the power spent in the transmitter. To carry out the optimization on T_d, the relation between wake-up radio power and NF should be known. As a simple topology-independent relation could not be found, a reasonable value of $T_d = 5.24\,\mu s$ has been chosen.

Effect of Interferers

Interferers can erroneously trigger the wake-up radio, thereby increasing the power consumption of the whole system. Due to the absence of a narrowband filter in the wake-up radio, any interferer in the antenna's bandwidth causes a false alarm with a probability dependent on the power of the interferer, exactly in the same way as the probability of missed detection is dependent on the power of the received signal. We take into consideration an interferer that has equal probability of generating a false alarm as the minimum signal detectable by the wake-up radio. This corresponds, for a non-pulse-based interferer to a power equal to

$$P_{int} = S \cdot CF = -74\,dBm \tag{2.25}$$

where S is the receiver sensitivity and CF is the Crest Factor defined in Sect. 2.4.2. It is clear that a gain equal to the Crest Factor is achieved in interference immunity thanks to the adoption of a pulse-based modulation scheme. In order to estimate the number of false alarms generated by interferers, we need to define an interferer scenario and the rate of appearance of such signals, which is not defined by spectrum regulations. As an example of a typical scenario, we consider the case of Bluetooth interferers caused by communication between a Bluetooth device, such as a mobile phone or a computer, and a Bluetooth headset. This is not sufficient to draw extremely general conclusions, but it allows us to sketch the activity of the main radio in a typical scenario. If any of the two Bluetooth devices is close enough to the node under consideration, a false alarm is issued when the Bluetooth device uses the channel. Bidirectional audio transmission between the headset and the other device typically generates traffic consisting of periodic repetition of bursts of two packets[13] (called SCO packets in the Bluetooth standard) with period[14] equal to 1.25, 2.5 or 3.75 ms [40]. In this scenario, the number of false alarms for each timeslot is, respectively, 160, 80 and 40 for the different repetition rates and the increase in system power consumption, due to increase of the duty-cycle of the main radio, is, respectively, 64, 32 and 16 μW.

In order to save energy in such a scenario, several strategies can be adopted. One strategy consists, for example, in limiting the number of wake-up requests for each timeslot. When, in a certain timeslot, the main radio is woken up by interferers more times than a certain maximum (defined as a parameter of the MAC protocol), it will not wake up anymore. Data transmission could then be delayed to other timeslots, during which the channel is not used by interferers. This strategy results in additional latency, which, however, can be tolerated in most applications. An alternative strategy consists in increasing the threshold of the comparator until the interferers can not trigger the wake-up radio. Increasing the threshold reduces the sensitivity and, consequently, the range between transmitter and receiver, i.e. d in (2.13). This strategy is particularly suitable for dense WSNs, in which information can be routed through multiple hops, in order to cover a longer distance with a sequence of shorter hops. The spatial redundancy of nodes typical of WSN can then be exploited to give robustness over interferers while maintaining low power consumption per node. Both strategies increase the packet latency and/or the packet rate, but this can always be accepted in low QoS applications. On the other hand, the system presented in this work could also implement applications with more stringent requirements on QoS. In that case, power consumption can be traded for QoS, by allowing interferers to wake up the main radio and, consequently, giving up the power saving gained by the introduction of the duty-cycled wake-up radio.

Notes on Implementation

Using the SNR found in the previous section and (2.22), the noise figure of the wake-up radio front-end can be derived. It is then possible to evaluate the feasibility of the proposed architecture in terms of power consumption. The noise bandwidth of the system is assumed to be 100 MHz, to allow spread both in the incoming frequency and in the center frequency of the receiver. For 100 MHz noise bandwidth and −11.8 dB signal-to-noise ratio, a noise figure of 21.6 dB is needed for a

---

[13]Each packet has a duration of 366 μs and the delay between the first and second packet in the burst is 625 μs.

[14]The period is adapted to the QoS required for the audio link.

sensitivity of −84 dBm. The envelope detector requires a non-linear device such as a diode. In this case, a minimum signal amplitude is required at the input of the envelope detector and, with reference to [38], we set the minimum amplitude to 60 mV$_{pp}$. Thus the voltage gain of the block preceding the envelope detector should have a voltage gain of 60 dB when an antenna impedance of 50 Ω and the parameters of Table 2.5 are used. Accordingly, the threshold for the comparator should be scaled to 144 mV. Taking into account similar architectures, it seems feasible to reach the specifications on gain and noise figure with 500 μW power consumption [18].

Moreover, ad-hoc low-power radio architectures can be developed, which exploits the properties of the adopted modulation scheme. This is for example shown in [41], where a wake-up radio for impulse radio signals with a power consumption of 415 μW is demonstrated. In that case, such a low power level can be reached at a high sensitivity of −82 dBm by adopting a conventional low-IF architecture and duty-cycling the receiver in such a way that the circuit is active only when IR pulses are expected.

2.5.3 Time Reference

The specifications for the time reference are based on the experimental results presented in Chap. 5. There, it will be shown that, with a power consumption less than 50 μW, a fully integrated time reference can reach an inaccuracy of the order of 1% over a wide temperature range, even when the effects of jitter and supply variations are taken into account.

2.6 Conclusions

This chapter has described how to implement a fully integrated node for Wireless Sensor Networks. The next generation of WSN will require very small and cheap nodes to be deployed by hundreds or thousands in the environment. To reduce both cost and size, each node must be fully integrated in a standard CMOS process and must be powered by energy scavengers or very small batteries. This limit on available energy sources poses a strong constraint on node power consumption, which must be lower than 100 μW. One of the main problems is then the integration of the node's frequency reference while maintaining a very low power consumption. Such a frequency reference is required both as a time reference to synchronize the network and as a reference for the wireless transceiver.

The time reference is used in a WSN to turn on the wireless transceiver only when a packet is expected and to keep the whole node in a low power sleep mode otherwise. As will be shown in later chapters, the accuracy of low-power fully integrated frequency references is much lower than that achieved by the usually adopted oscillators based on off-chip crystal resonators. An integrated time reference can not accurately predict packets arrivals and forces the transceiver to be active for a longer time to ensure data reception. To keep the power consumption below the $100 \, \mu W$ specifications, the *duty-cycled wake-up radio* approach was introduced. The time reference turns on a low-power wake-up radio when a packet is expected and the main radio is powered on demand only if an incoming packet is being transmitted. With this strategy, the inaccuracy required to the time reference is of the order of 1% for a total power consumption of the node of $73 \, \mu W$.

The RF section needs an accurate frequency reference in order to use the right portion of the wireless spectrum to communicate. An impulse-based modulation scheme has been adopted to relax the requirements on the frequency accuracy. It has been proven that a frequency inaccuracy of 1% is enough to respect spectral regulations and to enable tuning of the receiving nodes to the frequency used by the transmitter.

It has been shown that those techniques make the implementation of the required hardware blocks feasible. Thanks to the duty-cycled wake-up radio approach, the main radio is turned on only for packet transmission and reception, which represent a very small fraction of time for a low-activity WSN. Thus, its contribution to the total average power budget is marginal, leading to relaxed requirements for the main front-end design.

The adopted techniques also enable the implementation of a wake-up radio with an acceptable sensitivity. On the one hand, in fact, duty-cycling the wake-up radio relaxes the specification on its power consumption, and on the other hand the IR modulation is more immune to the effects of noise and interferers. In this way, the higher power budget and the robust modulation scheme can be exploited to achieve better performance.

Regarding the time reference, it can be concluded that a cheap low-power crystal-less WSN node can be realized as soon as a time reference with 1% inaccuracy and less than $50 \, \mu W$ of power consumption is available in a standard CMOS technology.

References

1. Weber W, Rabaey JM, Aarts E (eds) (2005) Ambient intelligence. Springer, New York
2. Akyildiz I, Su W, Sankarasubramaniam Y, Cayirci E (2002) A survey on sensor networks. IEEE Commun Mag 40(8):102–114. DOI 10.1109/MC.2002.1024422
3. Chong CY, Kumar SP (2003) Sensor networks: evolution, opportunities and challenges. Proc IEEE 91(8):1247–1256
4. EnOcean GmbH (2008) STM110 datasheet, Oberhaching, Germany, Sept. 2008. http://www. enocean.com. Accessed 4 Aug 2009
5. Crossbow Technology, Inc. (2009) XM2110CA datasheet, San Jose, CA. http://www.xbow. com. Accessed 4 Aug 2009
6. Dust Networks, Inc. (2009) M2600 datasheet, Hayward, CA. http://www.dustnetworks.com. Accessed 4 Aug 2009
7. Romer K, Mattern F (2004) The design space of wireless sensor networks. IEEE Wireless Comm 11(6):54–61. DOI 10.1109/MWC.2004.1368897

8. Weber W, Braun C, Dienstuhl J, Glaser R, Gsottberger Y, Knoll B, Lauterbach C, Leitner D,
   Shi M, Schnell M, Savio D, Stromberg G, Verbeck M (2005) Disappearing electronics and
   the return of the physical world. In: Symp. VLSI Circuits Dig. Tech. Papers, pp 45–48. DOI
   10.1109/VTSA.2005.1497076
9. Warneke B, Last M, Liebowitz B, Pister K (2001) Smart dust: communicating with a cubic-
   millimeter computer. Computer 34(1):44–51. DOI 10.1109/2.895117
10. Rabaey J, Ammer M, da Silva J JL, Patel D, Roundy S (2000) Picoradio supports ad hoc ultra-
    low power wireless networking. Computer 33(7):42–48. DOI 10.1109/2.869369
11. O KK, Kim K, Floyd BA, Mehta JL, Yoon H, Hung C-M, Bravo D, Dickson TO, Guo X,
    Li R, Trichy N, Caserta J, Bomstad WR II, Branch J, Yang D-J, Bohorquez J, Seok E,
    Gao L, Sugavanam A, Lin J-J, Chen J, Brewer JE (2005) On-chip antennas in silicon ICs and
    their application. IEEE Trans Electron Dev 52(7):1312–1323. DOI 10.1109/TED.2005.850668
12. Bryzek J, Roundy S, Bircumshaw B, Chung C, Castellino K, Stetter J, Vestel M (2006)
    Marvelous MEMS. IEEE Circ Dev Mag 22(2):8–28
13. Frank M, Kuhl M, Erdler G, Freund I, Manoli Y, Mller C, Reinecke H (2009) An integrated
    power supply system for low-power 3.3V electronics using on-chip polymer electrolyte
    membrane (PEM) fuel cells. In: ISSCC Dig. Tech. Papers, vol 1, pp 292–293
14. Al-Karaki J, Kamal A (2004) Routing techniques in wireless sensor networks: a survey. IEEE
    Wireless Comm 11(6):6–28. DOI 10.1109/MWC.2004.1368893
15. Epson Toyocom Corporation (2009) FC-12M datasheet, Tokyo, Japan. http://www.
    epsontoyocom.co.jp. Accessed 23 Aug 2009
16. SiTime Corporation (2009) SiT8003XT datasheet, Sunnyvale, CA. http://www.sitime.com.
    Accessed 23 Aug 2009
17. Discera Inc. (2009) DSC1018 datasheet, San Jose, CA. http://www.discera.com. Accessed 23
    Aug 2009
18. Daly DC, Chandrakasan AP (2007) An energy-efficient OOK transceiver for wireless sensor
    networks. IEEE J Solid State Circ (5):1003–1011
19. Pletcher N, Gambini S, Rabaey J (2007) A $65\mu W$, 1.9 GHz RF to digital baseband wakeup
    receiver for wireless sensor nodes. CICC Dig. Tech. Papers, pp 539–542
20. Le T, Mayaram K, Fiez T (2008) Efficient far-field radio frequency energy harvesting for
    passively powered sensor networks. IEEE J Solid State Circ 43(5):1287–1302. DOI 10.1109/
    JSSC.2008.920318
21. Yoo J, Yan L, Lee S, Kim Y, Kim H, Kim B, Yoo HJ (2009) A 5.2mW self-configured wearable
    body sensor network controller and a $12\mu W$ 54.9% efficiency wirelessly powered sensor for
    continuous health monitoring system. In: ISSCC Dig. Tech. Papers, pp 290–291
22. O'Driscoll S, Poon A, Meng T (2009) A mm-sized implantable power receiver with adaptive
    link compensation. In: ISSCC Dig. Tech. Papers, pp 294–295
23. Roundy S, Strasser M, Wright PK (2005) Powering ambient intelligent networks. In: Weber
    W, Rabaey JM, Aarts E (eds) Ambient intelligence. Springer, New York
24. Paradiso J, Starner T (2005) Energy scavenging for mobile and wireless electronics. IEEE
    Pervasive Comput 4(1):18–27. DOI 10.1109/MPRV.2005.9
25. Chalasani S, Conrad J (2008) A survey of energy harvesting sources for embedded systems.
    IEEE Southeastcon pp 442–447. DOI 10.1109/SECON.2008.4494336
26. Cook BW, Lanzisera S, Pister KSJ (2006) SoC issues for RF smart dust. Proc IEEE
    94(6):1177–1196
27. Ammer J, Burghardt F, Lin E, Otis B, Shah R, Sheets M, Rabaey JM (2005) Ultra low-power
    integrated wireless nodes for sensor and actuator networks. In: Weber W, Rabaey JM, Aarts E
    (eds) Ambient intelligence. Springer, New York
28. Rabaey JM, Ammer J, Karalar T, Suetfei L, Otis B, Sheets M, Tuan T (2002) PicoRadios
    for wireless sensor networks: the next challenge in ultra-low power design. In: ISSCC
    Dig. Tech. Papers, vol 1, pp 200–201
29. Enz CC, El-Hoiydi A, Decotignie JD, Peiris V (2004) Wisenet: an ultralow-power wireless
    sensor network solution. Computer, 37(8)

30. Pletcher N, Gambini S, Rabaey J (2009) A 52 μW wake-up receiver with -72 dBm sensitivity using an uncertain-IF architecture. IEEE J Solid State Circ 44(1):269–280. DOI 10.1109/JSSC.2008.2007438

31. Drago S, Sebastiano F, Breems L, Leenaerts D, Makinwa K, Nauta B (2009) Impulse-based scheme for crystal-less ULP radios. IEEE Trans Circ Syst I 56(5):1041–1052

32. Enz CC, Scolari N, Yodprasit U (2005) Ultra low-power radio design for wireless sensor networks. In: Proc. 2005 IEEE International Workshop on Radio-Frequency Integration Technology, pp 1–17

33. van Dam T, Langendoen K (2003) An adaptive energy-efficient MAC protocol for wireless sensor networks. In: Proc. First ACM Conference on Embedded Networked Sensor Systems (SenSys), pp 171–180

34. Ryckaert J, Desset C, Fort A, Badaroglu M, De Heyn V, Wambacq P, Van der Plas G, Donnay S, Van Poucke B, Gyselinckx B (2005) Ultra-wide-band transmitter for low-power wireless body area networks: design and evaluation. IEEE Trans Circ Syst I 52(12):2515–2525. DOI 10.1109/TCSI.2005.858187

35. Lee TH (2004) The design of CMOS radio-frequency integrated circuits, 2nd edn. Cambridge University Press, Cambridge

36. Hashemi H (1993) The indoor radio propagation channel. Proc IEEE 81(7):943–968. DOI 10.1109/5.231342

37. Couch LW II (1992) Digital and analog communication systems. Prentice Hall, NJ

38. Cook BW, Berny A, Molnar A, S L, Pister KSJ (2006a) Low-power 2.4-GHz transceiver with passive RX front-end and 400-mV supply. IEEE J Solid State Circ 41(12):2757–2766. DOI 10.1109/JSSC.2006.884801

39. Razavi B (1998) RF microelectronics. Prentice-Hall, NJ

40. Bluetooth Special Interest Group (2007) Specifications of the bluetooth system, Rev. 2.1, July 2007

41. Drago S, Leenaerts DMW, Sebastiano F, Breems LJ, Makinwa K, Nauta B (2010) A 2.4GHz 830pJ/bit duty-cycled wake-up receiver with −82dBm sensitivity for crystal-less wireless sensor nodes. In: ISSCC Dig. Tech. Papers, pp 224–225

Chapter 3
Fully Integrated Time References

What really matters anyway is not how we define *time, but how we measure it.*

Richard P. Feynman
The Feynman Lectures on Physics

3.1 Introduction

Measuring the time interval between two events requires the choice of a repetitive and regular phenomenon, such as the oscillation of a pendulum, and then counting how many times this phenomenon takes place between the two events. The science of timekeeping has evolved through the centuries by basically adopting more and more precise and reliable periodic phenomena to keep track of time. From the first attempts using evident astronomical events, such as the motion of the sun and the moon, chronometry evolved by employing periodic phenomena in man-made devices, such as sand motion in hourglasses, oscillations in pendulums, balance wheel rotations in mechanical clocks, electromechanical vibrations in quartz crystal oscillators and absorption or emission of radiations in atomic clocks.

Besides a periodic phenomenon, a clock requires two additional features [1]. First, the periodic motion must be sustained by feeding energy to the reference, for example pushing the weight of a pendulum to prevent friction from damping or even stopping the motion. The periodic phenomenon and the mechanism to feed it with energy constitutes an oscillator. Second, a counter is needed to keep track of the number of oscillations. While this is not a trivial problem for mechanical devices (solved only in the thirteenth century by the introduction of the escapement [2]), implementing a counter using microelectronic technology is straightforward. It is then clear that accurately measuring a time duration requires the availability of an oscillator which can produce a periodic output characterized by a stable and accurate frequency. For this reason, in the following, the expressions *frequency references* and *time references* will both be used to refer to such oscillators. The accuracy

F. Sebastiano et al., *Mobility-based Time References for Wireless Sensor Networks*,
Analog Circuits and Signal Processing, DOI 10.1007/978-1-4614-3483-2_3,
© Springer Science+Business Media New York 2013

of a frequency reference is determined by two main factors: the inherent accuracy of the periodic phenomenon, determined for example by its sensitivity to external conditions, such as the effect of temperature or external forces on a pendulum; and the particular implementation of the oscillator.

In this chapter, different periodic phenomena that can be used in fully integrated time references are reviewed and their practical implementations and relative limitations are discussed. First, the main sources of errors are discussed (Sect. 3.2). Then, after the detailed analysis of several kinds of time references, a comparison is presented (Sect. 3.9). As shown in Chap. 2, a fundamental building block for a WSN node is a low-power fully integrated time reference with moderate accuracy. Thus, in the benchmark, emphasis is given to those aspects that are crucial for such a time reference.

3.2 Sources of Errors

3.2.1 Basic Definitions

The output signal of an oscillator, commonly a voltage or a current, can be expressed as

$$s(t) = S(t)m[t + \epsilon(t)] \tag{3.1}$$

where $m(t)$ is a periodic signal with period T_0, $S(t)$ is the amplitude of the signal and $\epsilon(t)$ is the time error of the oscillator. Ideally, both $S(t)$ and $\epsilon(t)$ are constant and $s(t)$ is a perfect periodic waveform. However, perturbations will affect both the amplitude and the time error, causing distortion of the output signal. The first harmonic $x(t)$ of $s(t)$ can be expressed as[1]:

$$x(t) = A(t) \sin[\psi(t)] \tag{3.2}$$

$$\psi(t) = 2\pi f_0 t + \phi(t) + \theta(t) \tag{3.3}$$

where $\psi(t)$ is the instantaneous phase, $A(t)$ is the instantaneous amplitude and $f_0 = \frac{1}{T_0}$ is the oscillation frequency. The fluctuations of the instantaneous phase are split into two contributions: $\phi(t)$ are random variations; $\theta(t)$ are the systematic variations, due for example to temperature and aging. Both random and systematic perturbations affect also the amplitude $A(t)$. However, amplitude variations can easily be eliminated by, for example, limiting the amplitude of the signal. Moreover, amplitude variations do not affect the frequency and time accuracy of the oscillator

---

[1] $x(t)$ is obtained using the Fourier series expansion of $m(t)$, i.e. $A(t) = S(t)\sqrt{a_1^2 + b_1^2}$, $\phi(t) + \theta(t) = 2\pi f_0 \epsilon(t) + \arcsin \frac{a_1}{\sqrt{a_1^2 + b_1^2}}$, where $a_1 = \frac{2}{T} \int_0^{T_0} m(t) \cos(2\pi f_0 t) \, dt$, $b_1 = \frac{2}{T} \int_0^{T_0} m(t) \sin(2\pi f_0 t) \, dt$.

if $|A(t)| > 0$. As can be observed from (3.2), under that condition, $A(t)$ does not alter the phase of the oscillation and, for example, does not affect the instants of zero crossings, which can be used as repetitive events for time measurements.

The instantaneous frequency is defined as the derivative of the instantaneous phase:

$$f(t) \triangleq \frac{1}{2\pi}\frac{d\psi}{dt} = f_0 + \frac{1}{2\pi}\frac{d\phi}{dt} + \frac{1}{2\pi}\frac{d\theta}{dt} \tag{3.4}$$

Conversely, phase is the integral of frequency.

3.2.2   Random Errors

Frequency-Domain Characterization

Random errors in the oscillation frequency are due to noise in the physical components constituting the oscillator. They are described by the *phase noise* $\mathscr{L}(f)$ (pronounced "script-ell of f"), which is defined in [3] as

$$\mathscr{L}(f) \triangleq \frac{S_\phi(f)}{2} \tag{3.5}$$

where $S_\phi(f)$ is the single-sided Power Spectral Density (PSD) of the random phase fluctuations $\phi(t)$ [see (3.3)], expressed as a function of f, independent variable of the frequency domain. It can be proven that for small enough phase fluctuations, i.e. for $\int_0^{+\infty} S_\phi(f)df \ll 1 \, \text{rad}^2$, the measure of phase noise can also be expressed as

$$\mathscr{L}(f) \approx \frac{P_{sideband}(f, 1\,\text{Hz})}{P_{carrier}} \tag{3.6}$$

where $P_{sideband}(f, 1\,\text{Hz})$ is the power of one of the sidebands of the single-ended spectrum of the signal in a 1-Hz bandwidth at a frequency offset f from the carrier and $P_{carrier}$ is the power of the oscillation carrier [3]. From the last expression, it follows naturally that, when expressed in decibel, the units of $\mathscr{L}(f)$ are dBc/Hz, i.e. dB below the carrier in a 1 Hz bandwidth. Equation (3.6) gives a clear relation between the phase noise of the oscillator and the shape of the output spectrum.

The shape of $\mathscr{L}(f)$ depends on the power spectra of the noise produced in the oscillator. Typically, in an electronic oscillator, the random *frequency* fluctuations are directly proportional to the noise of active and passive devices. Since the phase is the integral of the instantaneous frequency, the PSD of the phase noise will have a slope of -20 dB per decade for white frequency noise and -30 dB per decade for flicker frequency noise, which in a typical case are caused, respectively, by thermal and flicker noise in the devices. An oscillator waveform affected by both flicker and

Fig. 3.1 Simulation of phase
noise in an oscillator in the
frequency and time domain:
(**a**) phase noise vs. frequency
offset from the oscillation
carrier; (**b**) long-term jitter
vs. time; extrapolated thermal
and flicker noise components
and their cumulative
contribution are also plotted

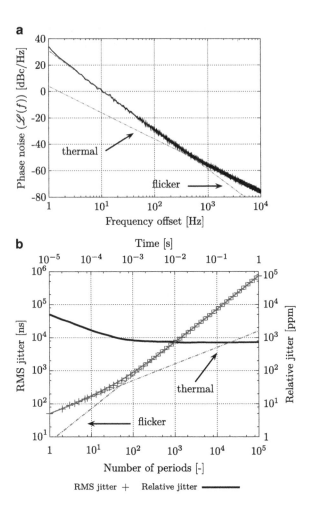

white frequency noise has been generated by simulations and its phase noise plot
with the two different slopes is shown in Fig. 3.1a.

The frequency-domain characterization of reference stability described above is
very useful when dealing with RF communication systems. For example, when
using the frequency reference as local oscillator for a down-conversion receiver, it
is straightforward to determine from a phase noise plot like the one in Fig. 3.1a the
amount of noise that can be down-converted by an interferer at a certain frequency
offset from the carrier [4].

Time-Domain Characterization

If the oscillator is employed in a time reference, it can be useful and practical to use
a time domain characterization of random errors: *jitter*. The jitter accumulated after

N clock periods is defined as the standard deviation of the cumulative duration of N successive clock periods. It can be expressed by considering the time between the zero-crossings t_1 and t_2 of $x(t)$ in (3.2) chosen such that

$$\begin{cases} \psi(t) = 2\pi f_0 t + \phi(t) = 0 \\ \psi(t + \tau) = 2\pi f_0(t + \tau) + \phi(t + \tau) = 2\pi N \end{cases} \tag{3.7}$$

The jitter $\sigma_j(\tau)$ accumulated after a time τ, where $\langle \tau \rangle = N T_0$, is defined as the standard deviation of the time error Δt:

$$\sigma_j(\tau) \triangleq \sqrt{\langle \Delta t^2 \rangle} \tag{3.8}$$

where

$$\Delta t = \tau - \langle \tau \rangle = \tau - N T_0 = \frac{\phi(t) - \phi(t + \tau)}{2\pi f_0} \tag{3.9}$$

where the last expression follows directly from (3.7). The relative jitter can be simply defined as

$$J_{rel}(\tau) \triangleq \frac{\sigma_j(\tau)}{\tau} \tag{3.10}$$

The jitter is often specified in terms of *period jitter*, i.e. the jitter in one period ($\sigma_j(\tau)$ for $\langle \tau \rangle = T_0$), and it is common to report it as *Root-Mean-Square (RMS)* or *peak-to-peak* value. In telecommunication systems, jitter is also measured in terms of *Unit Interval (UI)*, i.e. as a fraction of a period, which in communication systems refers to the duration of one of the symbols used in the modulation. When the quantity of interest is the jitter after a large number of periods, the term *long-term jitter* is used. In telecommunication systems, timing variations characterized by frequency components above or below $10\,\text{Hz}$ are defined, respectively, as *jitter* and *wander* [5]. Besides those more common time-domain jitter measurements, many other characterizations exist for jitter, depending on the application, and which are out of the scope of this work such as the *cycle-to-cycle jitter* (the duration difference between two adjacent periods) and the *N-cycle-to-cycle jitter* (the duration difference between two adjacent intervals each formed by N periods), which are of interest for digital circuit designers.

The jitter measure recommended by IEEE is the *two-sample variance*, or *Allan variance*, defined as [3]

$$\sigma_y(\tau) \triangleq \sqrt{\frac{1}{2} \left\langle [\bar{y}_2 - \bar{y}_1]^2 \right\rangle} \tag{3.11}$$

where $\bar{y}_k$ is the average fractional frequency given by

$$\bar{y}_k \triangleq \frac{1}{\tau} \int_{k \cdot T_0}^{k \cdot T_0 + \tau} \frac{f(t)}{f_0} dt = \frac{\phi(k \cdot T_0 + \tau) - \phi(k \cdot T_0)}{2\pi \tau f_0} \tag{3.12}$$

The main difference between the Allan variance and the standard variance defined in (3.8) and (3.9) is the presence in (3.11) of the difference between two adjacent samples of the frequency. It can be proven that, thanks to this characteristic, Allan variance is convergent for several kinds of noise, among which flicker noise, while the classical variance of (3.8) is not, as will be shown in the next section [6]. For this reason, Allan variance is used as a standard measure of frequency stability in high-precision oscillators, such as atomic clocks. However, such accurate analysis is not needed in this work and in the following the jitter defined in (3.8) and (3.10) will be used.

Relation Between Frequency-Domain and Time-Domain Representation

Using the relation between phase noise and time error expressed by (3.9), jitter can also be expressed in terms of the PSD of the phase noise[2] $S_\phi(f)$:

$$\sigma_j^2(\tau) = \int_0^{+\infty} S_{\Delta t}(f)\mathrm{d}f \tag{3.13}$$

$$= \frac{1}{(2\pi f_0)^2} \int_0^{+\infty} S_\phi(f) \left| 1 - e^{j2\pi f\tau} \right|^2 \mathrm{d}f \tag{3.14}$$

$$= \frac{T_0^2}{\pi^2} \int_0^{+\infty} S_\phi(f) \sin^2(\pi f\tau)\mathrm{d}f \tag{3.15}$$

where $S_{\Delta t}(f)$ is the PSD of Δt.

In case of white frequency noise, the phase noise PSD can be expressed as $S_\phi(f) = \frac{k_{white}}{f^2}$ and the white noise component of the jitter is given by[3]:

$$\sigma_{j,white}^2(\tau) = \frac{T_0^2}{\pi^2} \int_0^{+\infty} \frac{k_{white}}{f^2} \sin^2(\pi f\tau)\mathrm{d}f \tag{3.16}$$

$$= \frac{k_{white} T_0^2}{\pi} \tau \int_0^{+\infty} \frac{\sin^2 x}{x^2}\mathrm{d}x \tag{3.17}$$

$$= \frac{k_{white} T_0^2}{2} \tau \tag{3.18}$$

---

[2]Note that phase noise can in general also not be a stationary process, since its statistical property changes with time, such as, for example, its variance that can increase with time. The relation between its standard deviation and its PSD is then not well defined. However, defining jitter as a stationary process with parameter τ justifies the calculations presented in this section, even if not all of them are not formally correct, including the integration of $S_\phi(f)$ to compute the jitter.

[3]It can be proven, integrating by parts, that $\int_0^{+\infty} \frac{\sin^2 x}{x^2}\mathrm{d}x = \int_0^{+\infty} \frac{\sin x}{x}\mathrm{d}x$; the latter term can be proven to be equal to $\pi/2$ [7].

The same calculation can be executed for flicker noise, for which $S_\phi(f) = \frac{k_{flicker}}{f^3}$,

$$\sigma_{j,flicker}^2(\tau) = \frac{T_0^2}{\pi^2} \int_{f_l}^{+\infty} \frac{k_{flicker}}{f^3} \sin^2(\pi f \tau) df \tag{3.19}$$

$$= k_{flicker} T_0^2 \tau^2 \int_{\pi f_l \tau}^{+\infty} \frac{\sin^2 x}{x^3} dx \tag{3.20}$$

$$= k_{flicker} T_0^2 \gamma \tau^2 \tag{3.21}$$

where $\gamma = \int_{\pi f_l \tau}^{+\infty} \frac{\sin^2 x}{x^3} dx$ and f_l is the lower integration limit for the flicker noise. Note that the integral in (3.15) is divergent for flicker noise, due to the divergence of the spectrum of $1/f$ noise at low frequency. To solve this issue, which is related to the modeling of flicker noise, either a lower integration limit f_l must be used or a pole is introduced at low frequency in the flicker noise PSD [8,9]. In both cases, $\sigma_{j,flicker}$ results proportional to τ. Moreover, as observed in [10], noise at very low frequency is not easily distinguished from a drift, e.g. temperature drift.

As we can see from (3.18) to (3.21), the standard deviation of the jitter is proportional to the square root of the time in the short term and proportional to it in the long term. It is then easy to see that the relative jitter J_{rel} converges to a constant for long term when only flicker and white noise are present. This can also be observed in Fig. 3.1b, in which the jitter of the signal with the phase noise of Fig. 3.1a has been plotted. Since the relative jitter is the standard deviation of the relative error due to noise in the measurement of a time period, we can conclude that flicker noise is the most important factor for the design of frequency references used to measure relatively long time periods.

3.2.3 Systematic Errors

A parameter p, such as temperature or supply voltage, can cause systematic errors in the oscillator output frequency f. In case the sensitivity $\frac{\partial f}{\partial p}$ is a constant function of time, tracking of the parameter p allows the compensation of errors of f determined by p, in contrast to random errors which can not be compensated for. Typical parameters mostly affecting integrated circuits are PVT variations.

Process Spread

Due to imperfections in the IC manufacturing process, physical parameters, such as sheet resistance or oxide thickness, varies from die to die. Those variations directly affect the electrical properties of the circuit components, such as resistance or capacitance, which can show a typical spread up to 10–20% in a modern process [11]. If such a large variation affects a frequency reference, the resulting frequency error can be excessive for most applications. Since this error is static, it can be corrected by a single calibration.

The calibration can be performed at wafer level or after packaging. In wafer-level calibration, a wafer probe steps over the wafer, calibrating in fast sequence all the dies, while after-packaging calibration requires mounting and calibration of each sample separately. Moreover, if temperature strongly affects the oscillator frequency, the temperature at which the calibration is performed must be stabilized, accurately measured and taken into consideration in the calibration procedure. In wafer-level calibration, the temperature of the whole wafer, and consequently of the dies, is stabilized, while in after-packaging calibration the temperature of each sample, including the package, must be set, resulting in a longer and consequently more expensive fabrication.

However, after-packaging calibration can be required, since the same physical parameters can be subject to "packaging shift," i.e. a variation of their value due to the packaging process itself and which can clearly not be compensated at wafer level [12–14]. After dicing, the dies are glued with an epoxy adhesive to the leadframe, bondwires are attached between the die and the leadframe for electrical connection and plastic is molded around the die and the leadframe to form the final package. The adhesive and the plastic are processed at around 175°C and at this temperature the different materials, i.e. the silicon, the metal for the leadframe (usually copper) and the plastic, exhibit a different thermal expansion. After cooling at environment temperature, the die is then subject to mechanical stress of the order of 100 MPa, which can cause, for example, variations of a few percent in resistance of polysilicon or diffused resistors and in currents of MOS transistors [12]. Adoption of ceramic packages, which avoid plastic molding, or "sandwich" layers between the die and the plastic [13] can alleviate the effect of stress at the price of higher fabrication costs with respect to the use of cheaper plastic packages.

Temperature

Temperature variations affects the behavior of most components in an integrated circuit. The temperature range over which the proper operation of the frequency reference must be ensured, depends on the application. Most applications are specified over the military range (−55 to +125°C) in the industrial range (−40 to +85°C). If we take into consideration those ranges, variations can be as large as ±50%, as in the case of transistor parameters such as the carrier mobility, or smaller, of the order of few percents, as in the case of integrated resistors.

Variation of physical parameters will cause the frequency reference to show a finite temperature coefficient. If the temperature coefficient gives rise to variations larger than the required inaccuracy, additional compensation must be performed. This is possible if the temperature behavior of the reference is well defined and constant for all samples. In that case, a single calibration at room temperature is enough to trim out the process spread; then an integrated temperature sensor can measure the oscillator temperature and this information, together with the knowledge of the temperature coefficient, can be used to compensate the output frequency. Simpler compensation schemes avoid the use of a temperature sensor by compensating the output frequency with components showing temperature coefficients complementary to the temperature coefficient of the output frequency.

However, the temperature coefficient is always subject to process variations, since different samples show a slightly different temperature coefficient. This can be due to an intrinsic spread of the temperature coefficient of the physical phenomenon used as reference, or to the fact that different physical phenomena with different temperature coefficients define the frequency. In the latter case, different phenomena can dominate the temperature behavior in different temperature ranges and spread of the different physical phenomena causes random variation of the frequency temperature characteristic. If the spread of the temperature coefficient is large, temperature compensation based on a single-point trim can be inefficient and multi-point temperature trim must be employed. This procedure is however more expensive in terms of testing.

Supply Voltage

The sensitivity to variations of the supply voltage can be expressed either in terms of Power Supply Rejection Ratio (PSRR), i.e. the variation of the output normalized to the amplitude variation of the supply voltage vs. frequency, or supply pushing, i.e. the derivative of the output frequency as a function of the DC value of the supply voltage. PSRR is mostly used to quantify the error in case of interference coupled to the supply, such as mains interferences (50–60 Hz) or disturbances caused by other circuits connected to the same supply. Usually this is not a strong requirement for battery-powered systems. in contrast, supply pushing is more important, because the reference must operate within its specified accuracy even when the battery voltage diminishes as the battery charge is consumed.

Obtaining a reference weakly sensitive to supply variations is mostly achieved at the circuit design level. Usually the supply voltage does not directly affect the physical reference, but the oscillator circuit. It is then strongly dependent on the implementation and methods to prevent its effects will be presented in Chap. 4 devoted to circuit implementation.

3.3   RC-Based References

The time constant of an RC network can be used as the frequency-defining element of a fully integrated oscillator. The ultimate accuracy of such oscillators is determined both by the intrinsic accuracy of the resistance R and capacitance C, and by the particular implementation of the oscillator. In the following, first the intrinsic accuracy of the physical phenomenon is investigated, i.e. the accuracy of R and C, and then the possible implementations of the reference and the related design issues are discussed.

3.3.1 Frequency Accuracy

Accuracy of Integrated Resistors

Considering a thin-film resistor, the resistance R of a rectangular parallelepiped of length L, width W and thickness t, constituted by a material with resistivity ρ, is given by the well-known expression

$$R = 2R_{end} + \rho \frac{L}{Wt} = 2R_{end} + R_{sh} \frac{L}{W} \tag{3.22}$$

where R_{end} is the contact resistance at each end of the resistor and $R_{sh} = \rho/t$ is the sheet resistance. Variation in R can be minimized by increasing the resistance value, in order to make the first term in (3.22) negligible, and by increasing the resistor's physical dimensions, i.e. W and L, to reduce the effect of absolute errors in resistor geometry, due for example to lithography and etching. Furthermore, geometric errors also decrease with the shrinking feature size of new technologies. The main source of inaccuracy for integrated resistors is then the sheet resistance R_{sh}, which may vary by $\pm 20\%$ [11]. This large spread is due both to the spread in the resistivity and in the thickness. Note in fact that the thickness of any layer in an IC process can show variability of the order of $\pm 10\%$, while the errors in lateral dimensions, i.e. W and L, decrease with the shrinking feature size of the technology and are approximately independent of the size for large structures. Different resistor implementations are possible in integrated form, including diffused resistors, metal resistors, thin-film resistors and polysilicon resistors. Diffused resistors are often avoided if high precision is needed due to their excessive non-linearity[4] and leakage currents. Metal resistors show almost ideal properties, but their sheet resistance is of the order of $10 \, m\Omega/\square$ and is usually too small for RC oscillator applications. Thin-film resistors are fabricated using materials optimized to achieve very low temperature coefficients and high linearity, but they are not available in baseline digital processes. Polysilicon resistors are characterized by high linearity and sheet resistance ranging from $10 \, \Omega/\square$ to $1 \, k\Omega/\square$, depending on whether the low-resistivity polysilicon used for MOS gates is directly adopted or high-resistivity

---

[4]The linearity of passive components describes the variation of the component properties, i.e. resistance or capacitance, for a change in the voltage applied to the component. A high linearity is sign of robustness to voltage variation and consequently low sensitivity to supply voltage variations.

unsilicided polysilicon is used. These properties make poly resistors the preferred choice for RC-based references [15–17].

Apart from process spread, the main source of inaccuracy for poly resistors is their temperature coefficient, which can vary between $-2,000$ and $+1,000$ ppm/°C, depending on doping concentration, size of the silicon grains constituting the polysilicon and grain-to-grain boundary conditions [11, 18, 19]. To minimize the temperature coefficient of the output frequency of RC-based references, compensation of the temperature coefficient is often obtained by implementing a composite resistor using two different kind of resistors having opposite temperature coefficients. This can be achieved, for example, by using n-doped poly and p-doped poly, or the low-resistive poly and the high-resistive poly available in some technologies. The composite resistor than has a resistance equal to

$$R = R_p + R_n = R_{p0}(1 + \alpha_p T) + R_{n0}(1 + \alpha_n T) = (R_{p0} + R_{n0})(1 + \alpha T) \quad (3.23)$$

where $R_{p0,n0}$ are the extrapolated resistances at 0 K of the two resistors, $\alpha_{p,n}$ respectively their positive and negative temperature coefficient and α_{tot} the temperature coefficient of the composite resistor, which can be expressed as:

$$\alpha_{tot} = \frac{R_{p0}}{R_{p0} + R_{n0}}\alpha_p + \frac{R_{n0}}{R_{p0} + R_{n0}}\alpha_n \quad (3.24)$$

If $\alpha_{tot} = 0$, perfect compensation is achieved. Note, however, that the sheet resistance of R_p and R_n spreads in an uncorrelated way, since their fabrication steps are different. It is then interesting to investigate the effect of spread on the temperature coefficient α_{tot}. The error in α_{tot} due to the spread of R_{p0} (and likewise R_{n0}) is

$$\Delta\alpha_{tot,\Delta R_{p0}} = \frac{\partial \alpha_{tot}}{\partial R_{p0}} R_{p0} \frac{\Delta R_{p0}}{R_{p0}} = \frac{R_{p0} R_{n0}}{(R_{p0} + R_{n0})^2}(\alpha_p - \alpha_n)\frac{\Delta R_{p0}}{R_{p0}} \quad (3.25)$$

and using the condition $\frac{R_{p0}}{R_{n0}} = -\frac{\alpha_n}{\alpha_p}$, derived from (3.24) and under the assumption $\alpha_{tot} = 0$, and the inequality between the arithmetic and the geometric mean,

$$|\Delta\alpha_{tot,\Delta R_{p0}}| = \left|\frac{\alpha_n \alpha_p}{\alpha_n - \alpha_p}\right|\left|\frac{\Delta R_{p0}}{R_{p0}}\right| = \frac{|\alpha_n \alpha_p|}{|\alpha_n| + |\alpha_p|}\left|\frac{\Delta R_{p0}}{R_{p0}}\right| < \frac{|\alpha_n| + |\alpha_p|}{2}\left|\frac{\Delta R_{p0}}{R_{p0}}\right|$$

$$(3.26)$$

In a similar way, it can be proven that the error due to the spread of α_p (and correspondingly α_n) is

$$|\Delta\alpha_{tot,\Delta\alpha_p}| = \left|\frac{\partial \alpha_{tot}}{\partial \alpha_p}\alpha_p\frac{\Delta\alpha_p}{\alpha_p}\right| < \frac{|\alpha_n| + |\alpha_p|}{2}\left|\frac{\Delta\alpha_p}{\alpha_p}\right| \quad (3.27)$$

Taking into consideration a typical spread of the sheet resistance of $\pm 10\%$ and a typical temperature coefficient of around 1,000 ppm/°C, the resulting total temperature coefficient would be around 100 ppm/°C. Moreover, according to [11], it is difficult to hold temperature coefficients to tolerances of better than $\Delta \alpha_{p,n} \approx \pm 250$ ppm/°C. This will introduce an additional variation of the total temperature coefficient of $|\Delta \alpha_{tot,\Delta \alpha_p}| \approx 250$ ppm/°C. Thus, temperature coefficients of a few hundreds ppm/°C can be expected. However, as shown above, the residual temperature coefficient of a composite resistor is strongly dependent on the intrinsic temperature coefficients of the individual resistors (α_n, α_p). For some specific processes, temperature coefficients have been reported for poly resistors as low as ± 140 ppm/°C [15], which enables a temperature error of only about ± 14 ppm/°C.

Accuracy of Integrated Capacitors

Integrated capacitors can be implemented by using either "linear" capacitors, i.e. capacitors whose electrodes are fabricated using conductive layers, such as poly–poly capacitors or Metal-Insulator-Metal (MIM) capacitors, or device parasitic capacitance of active devices, such as MOS capacitors or junction capacitors.

Regarding process spread, junction capacitors exhibit spread as high as $\pm 20\%$ due to variations in doping concentrations and doping profiles [11]. Layer-to-layer and MOS capacitors can be approximated, at least locally, as parallel-plate capacitors, whose capacitance is given by

$$C = \epsilon_{die} \frac{WL}{t} \tag{3.28}$$

where ϵ_{die} is the permittivity of the dielectric (usually silicon dioxide and less frequently silicon nitride), W and L are the dimension of the plates and t is the thickness of the dielectric. Since three orthogonal dimensions[5] (W, L, t) concur in defining the capacitance, at least one of them is proportional to the thickness of a layer or of a set of layers in the IC. Consequently, any integrated capacitor, except junction capacitors, suffers from the thickness variability discussed in the previous section and its process spread will typically be about $\pm 10\%$.

The linearity of device capacitors is usually much lower than that of "linear" capacitors and particular care must be taken to ensure a constant voltage swing, regardlessly of supply voltage variations.

Among the "linear" capacitors, MIM capacitors are preferred because, unlike poly–poly capacitors, they can be built using the standard metal layers of a digital IC process. For MIM capacitors, with reference to (3.28), the temperature

---

[5]Reference capacitors can be constructed whose capacitance is only dependent on the permittivity and one dimension, such as the Thompson–Lampard capacitor [20, 21]. However, their implementation in an integrated circuit is unpractical.

Fig. 3.2 General block diagram of a linear *RC* oscillator

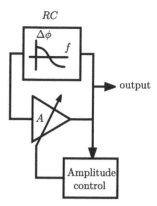

coefficient is determined by the temperature dependence of either the dielectric permittivity or the physical dimensions. The permittivity of silicon dioxide has a temperature coefficient of about 15–21 ppm/°C [22, 23]. The contribution due to thermal expansion of the dimensions is approximately 5.6 ppm/°C [22].

Of the device capacitors, MOS capacitors are usually preferred to junction capacitors because they are not as strongly affected by leakage currents. For MOS capacitors there is an additional effect due to the temperature variation of the charge density in the silicon: temperature coefficients ranging from 20 to 730 ppm/°C have been reported [22, 24, 25].

3.3.2 Implementation

Oscillators with output frequency proportional to $1/RC$ can be roughly classified in three categories:

- *Linear oscillators*, based on the fulfillment of the Barkhausen criterion and characterized by sinusoidal output, such as the classical Wien-bridge and phase-shift oscillators [26].
- *Non-linear oscillators*, based on the use of non-linear devices (Schmitt triggers, comparators), such as the astable multivibrator or the relaxation oscillator.
- *Locked oscillators*, in which a controlled oscillator, as for example a voltage-controlled ring oscillator, is inserted in a feedback loop in such a way that its oscillation period is forced to be proportional to an *RC* delay.

Linear oscillators are composed by three main building blocks (Fig. 3.2): the passive section, composed by resistors and capacitors, with a fixed phase-shift-vs.-frequency ($\Delta\phi$ vs. f) characteristic that fixes the oscillation frequency; a linear active section, i.e. an amplifier which provides energy and ensures that the Barkhausen condition for the amplitude is met; and an amplitude regulator, which limits the amplitude of the oscillation to prevent the non-linear operation of active

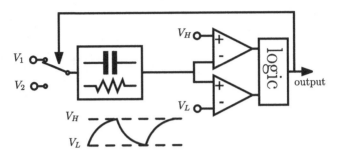

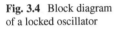

Fig. 3.3 General block diagram of a non-linear RC oscillator

Fig. 3.4 Block diagram
of a locked oscillator

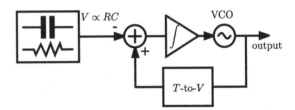

components and the consequent frequency errors. If the phase shift introduced by
the amplifier is negligible or, at least, constant, the oscillation frequency is fixed by
the passive section's phase-frequency relation according to the Barkhausen criterion
for the phase. Non-linearity of the amplifier affects the frequency of oscillation and
the amplitude control is employed to limit such effect. This block can, however,
strongly affect the temperature coefficient of the oscillator [15].

A typical block diagram of a non-linear oscillator is shown in Fig. 3.3. At the
start of the oscillation, the input of a passive RC filter is switched to V_1 and its
output shows a transient characterized by the time constant of the linear passive
filter. One of the two comparators detects the crossing of the voltage V_H and
an appropriate logic switches the input of the passive filter to V_2, causing the
filter output to approach the lower threshold V_L. The second comparator detect
the crossing and the input voltage is switched back to V_1, restarting the cycle.
The frequency of oscillation is fixed by the time constants of the RC filter and
can only be affected by delays introduced by the comparators and by their offsets.
Accuracy can then be granted by compensating the offsets and by using fast
switching comparators, without the need for accurate linear components or gain
control loops. The main drawback is the poor jitter performance of these oscillators,
caused by the noise introduced by the comparators [27].

In an RC locked oscillators (Fig. 3.4), a voltage proportional to the output period
of the VCO is produced by the T-to-V block and compared to a voltage proportional
to an RC delay. The frequency accuracy is determined by the precision in the
conversion between time and voltage and in the mechanism used to produce a
voltage proportional to RC. Different techniques can be employed to implement
such blocks (as in [16, 17, 28]) but the use of any additional circuit can spoil the

frequency accuracy, compared to the much simpler architectures of linear and non-linear oscillators. The main advantage is the use of an additional oscillator whose topology does not determine the frequency accuracy of the reference. In this way, some characteristics of the oscillator are decoupled from the accuracy: for example, low-jitter oscillators can be employed.

The power consumption of an RC oscillator is mainly determined by the current flowing through the resistors and by the consumption of the active blocks (amplifiers, comparators, VCOs). The first can be reduced by increasing the resistance and reducing the voltage swing across the resistors; the latter by lowering the frequency of operation. Although this could bring to a very low power, the jitter requirements poses a lower bound to the consumption [27], resulting in a consumption in the order of few microwatts in practical applications [15–17, 28].

3.3.3 Remarks

As discussed in the previous sections, the intrinsic frequency accuracy of RC oscillators is dependent on the accuracy of both resistors and capacitors. Both show a large process spread and consequently at least a trim at room temperature is required to obtain inaccuracy of the order of 1%. After trimming, their temperature dependence is the dominant source of inaccuracy, but variations of less than 1% can be obtained over a temperature range of more than 100°C [15–17]. This mainly depends on the availability of low-temperature-coefficient components in the adopted process. While MIM capacitors are easily available also in digital CMOS processes, resistors with very low temperature coefficients are not always available. Moreover, data on the spread of such temperature coefficients are usually not available and also the spread of the temperature coefficient of RC oscillators is often not available in published works [15–17].

3.4 LC-Based References

3.4.1 Frequency Accuracy

The basic form of a typical integrated LC oscillator is shown in Fig. 3.5a [29–32]. A parallel resonant tank is connected in parallel to an active circuit, whose equivalent resistance is negative and equal to $-1/g_{m,eq}$. The active circuit compensates for the losses of the tank and sustains the oscillation. In the tank, both the coil of inductance L and the capacitor of capacitance C exhibit finite losses, modeled respectively by series resistances R_L and R_C. Considering the linear system of Fig. 3.5a, the

Fig. 3.5 (**a**) Generic LC
oscillator block diagram.
(**b**) Typical CMOS
implementation (bias not
shown)

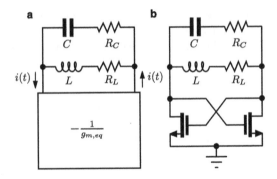

Barkhausen conditions for oscillation are satisfied if the active circuit compensates
for the losses and if the impedance of the tank shows a zero phase. The latter
condition is satisfied at frequency f, such that

$$f = f_0 \sqrt{\frac{L - CR_L^2}{L - CR_C^2}} \tag{3.29}$$

where f_0 is the tank's natural resonant frequency given by

$$f_0 = \frac{1}{2\pi\sqrt{LC}} \tag{3.30}$$

The oscillation frequency can also be expressed as

$$f \approx f_0 \sqrt{\frac{1 - \frac{CR_L^2}{L}}{1 - \frac{CR_C^2}{L}}} \approx f_0 \sqrt{\frac{1 - \frac{1}{Q_L^2}}{1 - \frac{1}{Q_C^2}}} \tag{3.31}$$

where Q_L and Q_C are respectively the inductor and the capacitor quality factors,
defined as

$$Q_L = \frac{2\pi f L}{R_L} \tag{3.32}$$

$$Q_C = \frac{1}{2\pi L R_C} \tag{3.33}$$

In practical cases, the oscillation frequency also depends on the implementation
of the active circuit. All parasitic capacitances and inductances of such circuit
can be embedded in the tank and (3.29) can be appropriately modified to include
their effect. However, (3.29) is only valid under the assumption of linearity of
the oscillator circuit and does not take into account that the circuit providing the
negative resistance will show non-linear behavior. The effect of the non-linearity can
be quantified by considering the harmonic content of the current $i(t)$ injected in the

tank (Fig. 3.5a). Under the assumption that the negative-resistance circuit does not store energy during the oscillation cycle,[6] the method of reactive power balance of harmonics [33] can be applied to find the difference between the effective oscillation frequency $f_{non-lin}$ and the oscillation frequency f given by (3.29):

$$\frac{f_{non-lin} - f}{f} = -\frac{1}{2Q_L^2} \sum_{n=2}^{+\infty} \frac{n^2}{n^2 - 1} h_n^2 \qquad (3.34)$$

where $h_n = \frac{I_n}{I_1}$ and I_n is the nth Fourier coefficient of the current $i(t)$. Beside the use of high-quality-factor inductors ($Q_L \gg 1$), the non-linear frequency deviation can be minimized either by reducing the harmonics of $i(t)$, for example by filtering them with passive filters [33], or by making the harmonic content constant with respect to PVT variations. The latter can be ensured by appropriate biasing of the active circuit to compensate the variations or by maximizing the harmonic content of $i(t)$ under any condition, i.e. by making $i(t)$ a square wave at the expense, however, of the increased power consumption needed to obtain a fast switching circuit under all PVT conditions.

If the effect of the non-linearity is minimized, the frequency of oscillation is determined only by the value of the tank's inductance and capacitance and their losses. The process spread of the oscillation frequency is mainly determined by the spread of integrated capacitors (already discussed in Sect. 3.3.1), which is of the order of 10% and much larger than the other sources of variations. The total effect of the losses accounts for a frequency variation of a few percent, as will be shown in the following, and its process spread is clearly smaller.

Regarding integrated inductors, they are implemented as planar coils using the thicker metal layers available in the adopted microelectronic technology. Their inductance is very weakly dependent on the thickness of the metal layers and the inductance value is mainly defined by the planar dimensions. Since inductances of the order of a few nH (typical in GHz-range applications) requires coil sizes of the order of $100\,\mu m$, the spread of the inductance is much smaller than the spread of the capacitance, considering that the lithographic inaccuracy of modern CMOS process is lower than $100\,nm$.

The tank inductance is also approximately independent of temperature variations [34], since those variations do not strongly affect the physical dimensions of the coil. The temperature dependence of the capacitance is the same as described in Sect. 3.3.1 and, for MIM capacitors, is mainly due to the variation of the dielectric permittivity. If the tank capacitor C is the only capacitive component in the tank, the temperature dependence of the capacitance directly affects the temperature

---

[6]This condition can be fulfilled by simply embedding in the oscillator tank all reactive components of the active circuit, such as the parasitic capacitances.

coefficient of the oscillation frequency and a compensation can be employed.[7] The losses in the tank (modeled by R_L and R_C) exhibit a strong variation with temperature. The capacitive losses can however be made negligible with respect to losses in the inductor by choosing an oscillation frequency low enough to ensure $Q_C \gg Q_L$ [see (3.32) and (3.33)]. In practice, this condition is fulfilled for typical oscillators in the GHz range. For such oscillators, besides the effects of the capacitance variation, the loss of the inductor represents the main source of inaccuracy.

The quality factor of integrated inductors is determined by two main factors: resistive loss and substrate loss. Resistive losses are dominant at low frequencies and are determined by the finite resistivity of the coil material. Aluminum is usually used for the metal layers of integrated circuits and its resistivity shows a positive temperature coefficient of 0.4%/°C. At higher frequency, the parasitic capacitance between the coil and the silicon substrate shows a lower impedance and currents flow in the substrate. The resistivity of the substrate has a positive temperature coefficient, but, since a higher resistivity hinders the flow of the current in the substrate, the total effect is a negative temperature coefficient of R_L at high frequency. In order to have a single physical effect, i.e. the temperature dependence of the resistivity of aluminum, dominating the temperature variation of R_L, the oscillation frequency must be kept low enough to minimize the substrate losses.

Neglecting capacitive effects, the temperature coefficient of the oscillation frequency can then be expressed as [31]

$$f_{TC} = \frac{1}{\omega}\frac{\partial \omega}{\partial T} = \frac{\omega}{\omega_0}\frac{1}{R_L}\frac{1}{Q_L^2}\frac{\partial R_L}{\partial T} \tag{3.35}$$

where $\frac{\partial R_L}{\partial T} = 0.4\%$ for aluminum. For example, with typical values $\omega \approx \omega_0 = 2\pi \cdot 1\,\text{GHz}$, $L = 1\,\text{nH}$, $Q_L = 10$, $R_L = 0.6\,\Omega$, $f_{TC} = 67\,\text{ppm/°C}$. From this expression, it is clear that a higher inductor quality factor decreases the temperature coefficient. A higher Q_L is also beneficial to minimize the effect of the active circuit's non-linearity, with reference to (3.34). The quality factor can be increased by adopting a higher oscillation frequency but an upper limit is fixed by the point where capacitive losses and inductor substrate losses become significant with respect to inductor resistive losses.

The temperature coefficient can be reduced also by appropriately compensating the temperature dependence of R_L by introducing ad-hoc capacitive losses R_C according to (3.29) [32].

---

[7]In order to fulfill this condition, the parasitic capacitances introduced by the active circuit must be minimized; those parasitic can be due to junction capacitances and MOS capacitances, which have different temperature coefficient than the tank capacitor and consequently can not be trimmed out by a temperature compensation.

3.4.2 Implementation

Power Consumption

While a trade-off exists in any oscillator between power consumption and phase noise, a minimum amount of power is always needed for the implementation of an oscillator, even without requirements on phase noise. This minimum power is usually of the order of a few microwatt or less for low frequency references such as RC oscillators, but it is not negligible for LC oscillators. In the following, a lower bound for the power required to LC references for oscillation is derived.

In the simplified circuit of Fig. 3.5a, power is spent in the active circuit to implement a negative resistance $-\frac{1}{g_{m,eq}}$ to compensate for the loss of the passive tank. If G is the real part of the admittance of the tank, we have, neglecting the loss in the capacitor ($R_C = 0$):

$$G = \mathbf{R}\left\{ j\omega C + \frac{1}{R_L + j\omega L} \right\} = \frac{R_L}{R_L^2 + \omega^2 L^2} = \frac{1}{R_L} \frac{1}{1 + Q_L^2} \qquad (3.36)$$

According to Barkhausen condition, a necessary condition for oscillation is

$$g_{m,eq} > G = \frac{1}{R_L} \frac{1}{1 + Q_L^2} \qquad (3.37)$$

In order to compute the power required to implement an admittance $-g_{m,eq}$, we assume the use of the circuit of Fig. 3.5b. This is a typical implementation of an LC oscillator in CMOS technology using a cross-coupled nMOS pair.[8] For this circuit, $g_{m,eq} = g_m$, where g_m is the transconductance of each of the two nMOS. In the following, we assume that the two nMOS are biased in weak inversion, which means

$$\frac{g_m}{I_D} = \frac{1}{n_{sub} V_t} \qquad (3.38)$$

where I_D is the drain current of each transistor, $V_t = \frac{kT}{q}$ is the thermal voltage and n_{sub} is the MOS subthreshold slope factor. This is the transistor region of operation where maximum g_m efficiency can be reached, i.e. the maximum g_m for a fixed drain current. Usually weak-inversion region is not employed for high frequency operation, since transconductance efficiency is paid in terms of reduced speed of the

---

[8]A more efficient circuit could be employed by adding a cross-coupled pMOS pair in parallel with the tank and re-using the same bias current of the nMOS pair. This would result in a factor 2 for the lower bound for the bias current but it would not radically change the conclusions.

devices, but the assumption is here useful to derive a lower bound for the current. The needed bias current is then

$$I = 2I_D = 2\left(\frac{g_m}{I_D}\right)^{-1} \quad g_m > 2n_{sub}V_t\frac{1}{R_L}\frac{1}{1+Q_L^2} \approx 2n_{sub}V_t\frac{1}{\omega L Q_L} \qquad (3.39)$$

where condition $Q_L^2 \gg 1$ has been used for the final approximation.

The following typical values are employed for the parameters in (3.39): $Q_L = 10$, $\omega = 2\pi \cdot 1\,\text{GHz}$, $n_{sub} = 1.5$ (typical values ranges between 1.2 and 1.6 [35]) and $L = 1\,\text{nH}$. With those parameters the minimum required current is $I = 1.2\,\text{mA}$. Taking into consideration that a higher power dissipation is usually required to ensure start-up of the oscillations under any PVT condition and that weak-inversion biasing can not be often adopted, it is clear that a power consumption of a few milliwatt is required even without considering any requirements on phase noise.

Design of the Integrated Inductor

As we have seen in Sect. 3.4.1, the loss of the integrated inductor is the main source of inaccuracy. Moreover, we can note in (3.39) in the previous section that the power consumption is inversely proportional to the inductance. It is then interesting to analyze the main design parameters of an integrated coil to understand the practical limits for Q_L and L.

An integrated inductor is realized by a metal conductor wound in a spiral with a number of loops or *turns*. The shape of the coils is usually circular, octagonal or square but it has a minor impact on the inductor performance. The major parameters that impact on the inductance are the number of turns n and the diameter of the coil. From basic physics and even from more accurate models [36], the inductance L is approximately proportional to the radius of the inductor and to n^2. Consequently, the inductance is limited by the area available on chip and, hence, by the cost. However, an increase in the dimensions of the inductors brings to an increase of the parasitic capacitance in parallel with the inductor. The frequency at which the inductance and the parasitic capacitance resonate is called the *autoresonance frequency* and it is the limit above which the inductor behaves like a capacitor and not anymore like an inductor [36]. Thus, the size of the inductor, and consequently the value of L and the power consumption [see condition (3.39)], are limited both by the area and the autoresonance. As will be shown in the following, R_L is proportional to L. This means that, with reference to (3.35), a larger inductance also decreases the temperature coefficient of the oscillation frequency.

The resistive losses of the inductor (and R_L) are proportional to the total length of the spiral and consequently proportional to the product of the radius and the number of turns. Since the inductance L is proportional to the product of n^2 and the radius, increasing the size of the coil does not increase the quality factor Q_L, while adding more turns is advantageous. However, it is not convenient to add many turns since the most internal windings contribute to the resistive loss in the same way but less to the inductance, since they are exposed to a smaller magnetic flux. In practice, hollow inductors featuring only two or three turns are usually employed. For a CMOS process, Q_L is then limited by the resistivity of the conductor and the thickness of the metal layers and it is usually smaller than ten [36].

3.4.3 Remarks

LC frequency references can be used as frequency or time references in practical circuits if a frequency divider is cascaded to the oscillator to generate a lower frequency signal from the GHz-range oscillator output. This division is beneficial in term of phase noise, since frequency division implies phase division and consequently attenuation of the phase noise. In this way, also if the Q of the tank is moderately high, its noise performance can be comparable with that of a low frequency resonator with much higher Q [31].

Those references have also been demonstrated to have a very low temperature coefficient (2 ppm/°C in [32]), mainly determined by the residual temperature coefficient of inductor losses after temperature compensation. However, a relatively large area for the integrated inductor and non-negligible power consumption even to start-up the oscillation are needed.

3.5 Thermal-Diffusivity-Based References

3.5.1 Principle

Electro-thermal systems are systems in which electrical variables (currents, voltages) are converted into thermal variables (temperature, heat) and vice-versa. Appropriate transducers facilitate the conversion: electrical heaters can be used to convert currents and voltages into heat, while temperature sensors generate electrical signals proportional to temperature. The simplified structure of an Electro-Thermal Filter (ETF) integrated on silicon is shown in Fig. 3.6: the filter input is the electrical power driven into a heater; the electrical power is dissipated as heat, which diffuses through the silicon substrate and generates a temperature gradient in the silicon; the temperature difference is sensed by a temperature sensor at a distance r from the heater and the temperature information is the filter output. Due to the thermal inertia of the silicon substrate, a finite time is needed for the heat to diffuse from the heater to the temperature sensor and this time can be adopted as a physical reference for a time/frequency reference.

Fig. 3.6 Simplified structure
of an electro-thermal filter
integrated on silicon

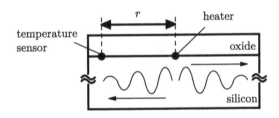

In the case of a punctiform heater and a punctiform temperature sensor, the
transfer characteristic of the filter can be expressed as [37]

$$H(\omega) = \frac{T_{sens}(\omega)}{P_{heat}(\omega)} = \frac{1}{2\pi k r} \exp\left(-r\sqrt{\frac{\omega}{2D}}\right) \exp\left(-jr\sqrt{\frac{\omega}{2D}}\right) \qquad (3.40)$$

where T_{sens} is the temperature reading of the temperature sensor, P_{heat} is the
power dissipated by the heater, k is the thermal conductivity of the substrate and
D is the thermal diffusivity of the silicon. If the filter is used in an oscillator
as the frequency-defining element, the phase-frequency characteristic of the filter
will fix the oscillation frequency. The oscillation frequency f will be such that the
phase shift between input and output of the ETF is equal to a reference phase ϕ_{ref}
dependent on the chosen topology for the oscillator, i.e.

$$\angle H(\omega) = \phi_{ref} \Rightarrow f = \frac{\omega}{2\pi} = -\frac{D\phi_{ref}^2}{\pi r^2} \qquad (3.41)$$

The frequency accuracy then depends on the spread of the heater-sensor distance
r and the die thermal diffusivity D. For this reason, these references are called
Thermal-Diffusivity-based (TD-based) references.

3.5.2   Frequency Accuracy

The accuracy of TD-based frequency references is limited by the accuracy of the
factors in (3.41). The relative frequency error is given by

$$\frac{\Delta f}{f} \approx \frac{\Delta D}{D} - 2\frac{\Delta r}{r} \qquad (3.42)$$

The spread of the thermal diffusivity of silicon is practically zero for the high-purity
low-doping silicon substrates employed in modern CMOS technologies and can be
neglected. However, though well defined, the temperature dependence of silicon
diffusivity is very strong and can be approximated in the industrial temperature
range as $\sim T^{-n}$ with $n \approx 1.8$ [37]. Consequently, the temperature of the die
must be accurately monitored by an on-chip temperature sensor and temperature

Fig. 3.7 Simplified block diagram of an TD-based frequency reference

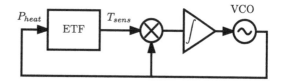

compensation must be performed. However, the error introduced by a practical temperature compensation can not be neglected. As shown in [38], a 0.1°C error in temperature sensing[9] causes a frequency error up to approximately 0.05%, due to the strong temperature dependence of thermal diffusivity.

The other important source of error is the spread of the heater-sensor distance Δr, which is determined by the lithographic accuracy of the process and can be assumed to be in the order of 10% of the minimum feature size of the process. The relative error $\frac{\Delta r}{r}$ can be reduced by increasing r but at the cost of reducing the output of the ETF. As can be seen from (3.40), the output of the ETF quickly diminishes with increasing distance, degrading the signal-to-noise ratio and hence the noise performance of the oscillator. An increased r could be compensated by increasing the power at the heater, but, beside the increase in power consumption, the heater power is limited by self-heating of the substrate. With a local increase of the temperature of the silicon, the diffusivity would change according to the T^{-n} law and could not be anymore correctly compensated by the on-chip temperature sensor, which can not track the local temperature variation at the ETF. With reference to the ETF described in [38], a heater-sensor distance of a few tens of μm is needed to implement an TD-based reference with an inaccuracy of less than 0.1% in a 0.7-μm CMOS technology ($\Delta r \approx 70$ nm). For such an ETF, a heater power of 10 mW is required to generate an ETF output of the order of only 1 mV. However, the accuracy of TD-based references benefits from Moore's law thanks to its dependency from lithographic accuracy, as shown in [40].

3.5.3 Implementation

Different architectures have been proposed to generate a frequency locked to the silicon diffusivity according to (3.41) [37, 38, 41, 42]. Those architectures are examples of the locked oscillators already treated in Sect. 3.3.2 and their simplified block diagram is shown in Fig. 3.7. The output signal of a VCO is sent to an ETF and the ETF output is mixed with the VCO output signal. The mixer operates as a phase detector, generating an output whose average depends on the phase shift between the ETF input and output. The loop is closed in a negative feedback loop

---

[9]State-of-the-art CMOS temperature sensors can achieve an inaccuracy of the order of 0.1°C, as shown in [39] and as will be further discussed in Chap. 5.

through an integrator which filters the mixer output and drives the control voltage of the VCO. The loop will reach the steady state when the input of the integrator has a zero average, or equivalently when the two inputs of the mixer are in quadrature. This is equivalent to locking to the frequency in (3.41) for $\phi_{ref} = -\pi/2$.

As already discussed in Sect. 3.3.2, the main advantage of this architecture is the possibility to limit the bandwidth of the loop by controlling the integrator bandwidth. This is a crucial factor for TD-based circuits, since the voltage at the output of the ETF is of the order of only 1 mV and noise must be filtered to achieve reasonable jitter performance. In a 0.7-μm CMOS process, the noise bandwidth has been limited to 0.5 Hz in [37] by the use of a large external capacitor in the integrator, while a fully integrated solution is adopted in [38], which achieves the same noise bandwidth by inserting a digital low pass filter between the integrator and the VCO. The drawback of such a low loop bandwidth is a slow response to variation of external parameters, such as temperature or supply voltage, with a consequent temporary drift of the frequency from its nominal value. The scaling of the ETF in a more advanced technology (0.16-μm CMOS in [42]) and the related decrease of the jitter allows for a noise bandwidth an order of magnitude wider, alleviating the issue.

One of the main practical drawback of TD-based references is the need of high precision circuits to handle the very-low-level signals at the output of the ETF. Although the feasibility of such low-noise low-offset front-ends has been proven [38, 41], it strongly affects the complexity of the design.

3.5.4 Remarks

TD-based frequency references have been proven to achieve a spread after trimming of less than 0.1% in 0.7-μm CMOS and in 0.16-μm CMOS [38, 42]. The spread is in that case limited by the lithographic accuracy and a better accuracy is expected when adopting a scaled CMOS technology. It is foreseen in [43] that with a scaled technology the inaccuracy will be limited to 0.05% (3σ) by the accuracy of the temperature sensor needed for compensating the $T^{-1.8}$ behavior of silicon diffusivity.

Technology scaling would also enable a scaling of the power needed at the heater: for the same accuracy, the distance between heater and sensor r can decrease by the scaling factor and keeping the amplitude of the ETF output constant, the heater power could decrease. However, even scaling the ETF proposed in [38] from 0.7-μm CMOS to 65-nm CMOS, the heater power would scale down from 2.5 mW to about 250 μW, which can be still limiting for some applications.

3.6 MOS-Based References

3.6.1 Principle

Various oscillator architectures can be built using only MOS transistors and capacitors, such as ring oscillators [44] or astable multivibrators [45]. The reference phenomenon is the charge (or discharge) of a capacitor between two reference voltages by the drain current of a MOS transistor. In general, the oscillation period can be approximated as the weighted sum of capacitors charge/discharge times and the frequency can be written as

$$f = \left(\sum_i \alpha^{(i)} \frac{I_D^{(i)}}{C^{(i)}(V_{r2}^{(i)} - V_{r1}^{(i)})} \right)^{-1} = \left(\sum_i \alpha^{(i)} \frac{\frac{1}{2}\mu^{(i)} C_{ox} \frac{W^{(i)}}{L^{(i)}} (V_{gs}^{(i)} - V_t^{(i)})^2}{C^{(i)}(V_{r2}^{(i)} - V_{r1}^{(i)})} \right)^{-1}$$

(3.43)

where the index (i) refers to the ith charge/discharge time and I_D is the drain current of a MOS transistor in saturation, α the weight factor, V_{r1} and V_{r2} reference voltages, μ_n is the carrier mobility, C_{ox} the oxide capacitance, W and L the transistor sizes, V_t the MOS threshold voltage and V_{gs} the gate-source voltage.

3.6.2 Frequency Accuracy

The individual terms in (3.43) will all exhibit errors which will contribute to the total inaccuracy. Reference voltages can be very accurately generated in an integrated circuit by exploiting the bandgap principle [10]: inaccuracy lower than 0.15% is achieved over the industrial temperature range by state-of-the-art voltage references [46]. Thus, the limit on the frequency accuracy of MOS-based oscillators is usually determined by capacitor accuracy (described in Sect. 3.3.1) and MOS current accuracy. The latter is determined by the bias point chosen for the transistor, i.e. gate, drain and source voltages, which can also be accurately set using bandgap references, and by the physical parameters of the transistors, i.e. carrier mobility, size (width and length) and threshold voltage.

Transistor size is strongly insensitive to environmental variations, but it can spread due to lithographic inaccuracy. However, by enlarging the device itself, the relative error on the device size can be minimized. The major contributors to inaccuracy are then the mobility and the threshold voltage, which are strongly process and temperature dependent. Mobility is approximately proportional to T^{-n} with $n \approx 1.5$, has a process spread of a few percent and is described in detail in Sect. 3.7. Threshold voltage has a typical temperature coefficient around $-2\,\text{mV/K}$. It is of the order of few hundreds of mV in modern CMOS processes and shows process spread of few tens of millivolt ($\approx$5–10%).

Thus, the process spread is of the order of tens of percent and a single-point trim is needed for most applications. The temperature coefficient is usually dominated by the effects of carrier mobility and threshold voltage. Even if their temperature behavior would be perfectly known and a perfect theoretical compensation could be applied separately to each of them, (3.43) shows that the oscillation frequency is a complex function of threshold voltages and carrier mobility of both nMOS or pMOS transistors. The combination of different temperature coefficients and different spreads of the various physical parameters produces a temperature coefficient and a spread of the oscillation frequency which can not be predicted and consequently compensated without an expensive multi-point temperature trimming scheme. To address this issue, the reference voltages (V_{r1} and V_{r2}) and the bias point of the transistors (determined for example by V_{gs}) should be generated with the appropriate value and temperature dependence to compensate for variation of threshold voltage and mobility.

3.6.3 Implementation

The basic element of MOS-based oscillators is a delay cell whose delay is proportional to the charge/discharge time of capacitors, as described in (3.43). Ideally, reference voltages and the bias point of MOS transistors inside the delay cell can be separately chosen and optimized to compensate for both carrier mobility and threshold voltage variations. However, in the typical implementation the delay cell is provided with only one control knob, to compensate for variations. The delay cells are then arranged in oscillators configurations, such as ring oscillators [44] or astable multivibrators [45].

The use of only one control point per delay cell, usually a control voltage, makes the compensation of the different spreads and temperature dependence not very accurate. Complex circuits are needed to produce the appropriate behavior for the control voltage and this can be accomplished at design time only if very accurate models for the temperature coefficients and the spread of the physical parameters are available.

Notwithstanding the complexity of this one-knob compensation approach, good results can be reached. In [44], a differential ring oscillator is compensated using only one control voltage and, by adoption of the appropriate biasing, the frequency variation is less than 2.6% over multiple runs and a large temperature range.

3.6.4 Remarks

The design of MOS-based oscillators usually consists in building an oscillator based on a simple topology, such as a ring oscillator, and then trying to compensate for its PVT variations with a complex biasing circuit. The drawback is that the oscillation frequency is not clearly locked to a single physical phenomenon and the compensation for temperature and process spread strongly relies on the availability of reliable models of integrated components. Besides the fact that such models are not always available for any digital CMOS technology, this limits the reliability and the portability to different technologies.

3.7 Mobility-Based References

3.7.1 Principle

Mobility-based references are a subset of MOS-based references. The main idea, illustrated in Fig. 3.8, is to bias a MOS transistor in such a way that its drain current is independent of the threshold voltage V_t [47, 48]. This current is used to discharge (or charge) a capacitor between two reference voltages V_{r1} to voltage V_{r2}. Its frequency is given by

$$f = \frac{I_D}{C(V_{r1} - V_{r2})} = \frac{\frac{1}{2}\mu_n C_{ox} \frac{W_1}{L_1}(V_{gs} - V_t)^2}{C(V_{r1} - V_{r2})} = \mu_n \frac{W_1}{2L_1} \frac{C_{ox}}{C} \frac{V_R^2}{V_{r2} - V_{r1}} \quad (3.44)$$

where V_R is a reference voltage.

3.7.2 Frequency Accuracy

Equation (3.44) express the dependency of frequency on other physical parameters that constitute also the main sources of inaccuracy, i.e. charge mobility, capacitances and reference voltages. As for MOS-based references, reference voltages can be generated with high absolute accuracy and a low temperature coefficient [46]. If all reference voltages (V_0, V_{r1}, V_{r2}) are derived from a common bandgap reference by voltage division or multiplication, we can assume that their effect on total inaccuracy

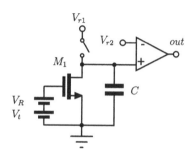

Fig. 3.8 Principle of mobility-based time reference

is much lower than 1% due to their absolute spread and temperature coefficient. As discussed in Sect. 3.3.1, capacitors shows a large spread but their temperature coefficient is negligible compared to that of charge mobility, which described in the following.

Charge mobility in MOS transistors is determined by different physical scattering mechanisms to which the charge are subject in the MOS channel, i.e. phonon scattering, Coulomb scattering and surface roughness scattering [49]. It is characterized by a low process spread of the order of few percent [47]. Mobility is strongly dependant on the bias condition of the MOS transistors, particularly on the gate-source voltage V_{GS}, since different scattering effects dominates the mobility behavior for different values of the transversal electric field in the channel [49]. Thus, care must be taken to ensure stabilization of the voltages at the MOS terminals. Although the temperature dependence of the mobility is large (approximately $T^{-1.5}$), it is well defined and can be compensated for.

3.7.3 Implementation

Delay cells or time references implementing the mobility-based principle are typically based on structures similar to a relaxation oscillator, i.e. charge or discharge of capacitors by means of a constant current [47, 48]. The current is in this case obtained by compensating the threshold voltage of a MOS transistor and the temperature dependence of the mobility.

Both in [47] and [48], the threshold voltage cancellation is obtaining by generating the voltage source V_t of Fig. 3.8 as a copy of the gate-source voltage of a MOS transistor with a negligible overdrive voltage. This method provides a rough compensation of the threshold voltage but it is inaccurate since it relies on the definition of the threshold voltage as the gate-source voltage for an arbitrary low current density in the MOS transistor. The latter is an inaccurate definition due to the transition, at low current densities, between the strong-inversion and weak-inversion modes of operation.

Assuming ideal compensation of the threshold voltage, the temperature behavior of the charge mobility is obtained by appropriately adapting the temperature dependence of V_R in (3.44). This can be done by obtaining V_R as sum of base-emitter voltages on bipolar transistors (which have a negative temperature coefficient) [48] or by taking V_0 as the voltage drop across a resistor biased by a combination of currents proportional to absolute temperature and proportional to the square of absolute temperature [47].

3.7.4 Remarks

An inaccuracy lower than $\pm 1\%$ has been achieved for one sample over a temperature range larger than 100°C, proving some effectiveness of the temperature stability of this principle [47]. Statistical data proving the stability of such a reference over process spread are reported in the following chapter. For further discussions the reader is referred to Chap. 4.

3.8   Other References

The frequency references described in the previous sections can all be fully integrated in a standard CMOS process. They exploit the electrical parameters of microelectronic components, such as resistance, capacitance or mobility, and in one case, i.e. for ETF, the thermal properties of the silicon substrate. However, additional processing steps in the standard CMOS process flow allow the fabrication of frequency references based on the mechanical properties of silicon. This can be accomplished by MEMSs, which are transducer between electrical and mechanical domain.

In particular, vibrating microresonators, can be used as frequency-defining elements in integrated oscillators, in which the output frequency is locked to the mechanical resonance frequency of the micromachined structure [50]. Very high quality factors, comparable to the one reached by quartz crystal, can be achieved, resulting in a lower power consumption compared to fully integrated solutions. Moreover, the absolute value of the resonance frequency is usually determined by the lateral dimensions of the structures and is consequently very accurate, taking advantage of the accuracy in defining such dimensions in microelectronic processing. Their temperature dependence is usually dominated by the temperature variations of the elastic properties of silicon, which exhibit temperature coefficients in the order of -30 ppm/°C. However, it is believed that such temperature coefficients are highly reproducible [51], thanks to the high purity and low defect density of the silicon employed for CMOS manufacturing, similarly to what happens in TD-based references (see Sect. 3.5.2). The temperature variations could then be compensated using the appropriate temperature compensation scheme.

By virtue of those properties, frequency references achieve a ± 10 ppm inaccuracy over a 50°C temperature range after a three-point temperature calibration with power consumption of a few microwatts [51] and a ± 30 ppm inaccuracy over the temperature range from -40 to 85°C after a single temperature calibration with power consumption of a few milliwatts [52]. In addition to those, commercial products with inaccuracy less than 100 ppm have been proposed as a low-cost alternative to quartz crystals [53, 54].

Despite the appealing performance, such devices can be manufactured only by adding processing steps to a standard microelectronic technology, if they need to be fully integrated. While this would increase the final cost of the frequency reference, they are also out of the scope of the present discussion, whose focus is on fully integrated frequency references in *standard* microelectronic processes.

3.9 Benchmark

3.9.1 Requirements for Time References for WSN

In Chap. 2, the requirements for a time reference for WSNs have been drawn. Those specifications are here reviewed under the light of the error sources discussion presented in this chapter.

The inaccuracy of a WSN time reference has to be lower than 1% taking into consideration both noise and environmental (PVT) variations. With regards to noise, since the clock on board of a WSN node is used to measure time periods ranging from hundreds of milliseconds to even some seconds (see Chap. 2), flicker frequency noise is usually the dominant source of error (Sect. 3.2.2). Moreover, note that in the case of flicker and white noise the assumptions about reference accuracy made in Chap. 2 are valid. It was there assumed that for the long term, the error of a time reference is directly proportional to the duration of the measured period [see (2.1)].

Requirements on temperature stability of the reference are fundamentals since in many applications WSN nodes are subject to a wide range of temperature variations. Accurate temperature compensation can be achieved by means of multi-point trimming schemes. This procedure is however more expensive in terms of testing and is not compatible with the low-cost requirements. Consequently, the number of temperature points at which calibration is performed must be minimized. At least a single-point trim is necessary when using most of the reference principles presented in this chapter, which can not achieve an inaccuracy below 1% due to the large process spread.

PSRR is usually not a strong requirement for WSN references since they are battery powered and operate when most of the on-board circuitry is in sleep mode and does not inject noise over the supply lines. Supply pushing is a fundamental limit, because the reference must operate within its specified accuracy during the whole WSN node lifetime, i.e. even when the battery voltage diminishes as the battery charge is consumed for battery-powered nodes, or if the energy source is unable to provide the same supply voltage at all times for energy-scavengers nodes.

The time reference must reach the 1%-accuracy level under those condition while dissipating less than $50\,\mu$W and occupying the least possible die area to reduce costs.

3.9.2 Discussion

The experimental performance of the different reference principle are compared in Table 3.1. The data, extracted from state-of-the-art works from the open literature, include RC oscillators, an LC oscillator, an TD-based oscillator and MOS-based oscillators (ring oscillators and a multivibrator). Mobility-based oscillators are not included in the comparison because no significant data are available. A fair

Table 3.1 Benchmark of frequency references

	[15]	[16]	[17]	[28]
Reference principle	RC	RC	RC	RC
Frequency	6 MHz	10 MHz	30 MHz	14 MHz
Temperature range	0–120°C	−20–100°C	−20–100°C	−40–125°C
Supply voltage	1.2 V	1.2 V	1.8 V	1.8 V
Power	66 μW	80 μW	180 μW	45 μW
Area	0.03 mm^2	0.22 mm^2	0.08 mm^2	0.04 mm^2
Process	65-nm	0.18-μm	0.18-μm	0.18-μm
Temperature variation	±0.5%	±0.4%	−0.7–0.5%	±0.19%
Temperature coefficient	86 ppm/°C	58 ppm/°C	100 ppm/°C	23 ppm/°C
Supply sensitivity	N.A.	0.03%/V	4%/V	0.8%/V
Noise Phase noise @ 1 MHz	−96.6 dBc/Hz @ 100 kHz	N.A.	−96 dBc/Hz	−116 dBc/Hz
Period jitter	N.A.	N.A.	N.A.	30 ps$_{rms}$
Process sensitivity	N.A.	1 sample	N.A.	N.A.
	7 samples		20 samples	1 sample

(continued)

Table 3.1 (continued)

	[32]	[42]	[44]	[45]	[55]
Reference principle	LC	TD	MOS (ring)	MOS (multivib.)	MOS (ring)
Frequency	25 MHz	16 MHz	7 MHz	680 kHz	680 kHz
Temperature range	0–70°C	−55–125°C	−40–125°C	−35–85°C	35–115°C
Supply voltage	3.3 V	1.8 V	2.5 V	1.8 V	4 V
Power	6.6 mW	2.1 mW	1.5 mW	3 μW	400 μW
Area	~1 mm^{2a}	0.5 mm^2	N.A.	0.015 mm^2	0.075 mm^2
Process	0.13-μm	0.16-μm	0.25-μm	0.13-μm	0.6-μm
Temperature variation	±80 ppm	±0.1%	±0.84%	±1%	±0.85%
Temperature coefficient	2 ppm/°C	±11.2 ppm/°C	102 ppm/°C	83 ppm/°C	106 ppm/°C
Supply sensitivity	0.001%/V	N.A.	0.88%/V	1.1%/V	N.A.
Noise Phase noise @1 MHz	−154 dBc/Hz	N.A.	N.A.	−105 dBc/Hz	N.A.
Period jitter	2 ps$_{rms}$	45 ps$_{rms}$	N.A.	2 ns$_{rms}$	N.A.
Process sensitivity	N.A.	<0.4%	<2.12%	N.A.	<4.7%
	32 samples	24 samples	94 samples	1 sample	29 samples
			4 batches		2 batches

^{a}Estimated

comparison is not always possible, especially due to the lack of experimental data extracted from a significant number of samples. However, some remarks can be drawn, taking also into consideration the previous analysis.

It appears clear that high accuracy, especially in terms of temperature stability, can be achieved by adopting LC references or TD-based references. In both cases, the inaccuracy is lower than 0.1% after a single-point trim and only limited to $\pm 0.2\%$ without any trimming for the TD-based one. At the same time, the noise performance of the LC reference is clearly better, thanks to the quality factor of integrated inductors and to the low signal-to-noise ratio at the output of typical ETF filters. However, those performance are paid in terms of power, which, as shown also in Sects. 3.4.2 and 3.5.2, can not practically be lower than few hundreds microwatt, even if state-of-the-art deep-submicron technologies would be adopted. Mainly for this reason, those references can not be adopted for ultra-low-power WSN nodes. Moreover, LC references embed an integrated inductor occupying a large silicon area that can hardly be reduced, with a consequent unavoidable cost.

On the other hand, RC oscillators and MOS-based oscillators require a power consumption down to the order of 100 μW while achieving inaccuracy over temperature in the order of 1%. Moreover, a small area on silicon is usually needed for such oscillators. While meeting the requirements for WSNs with those characteristics, their practical use could be limited for different reasons. The accuracy of RC oscillators is limited by the availability of resistors with low, or at least well-defined and characterized, temperature coefficient (see Sect. 3.3.1). Since this condition is strongly dependent on the technology, the adoption of such references can be limited for digital CMOS processes.

A similar argument holds for MOS-based references. The accuracy of such oscillators is determined by the mutual compensation of different physical effects. Thus, accurate oscillators can be obtained only if accurate models for the technology are available or additional runs are employed to fully characterize the oscillators. Hence, the portability and reliability of such references is not ensured.

The current state-of-the-art is summarized in Fig. 3.9, in which the accuracy and the power consumption of the frequency references of Table 3.1 are reported. Several RC and MOS-based references show inaccuracy and power consumption low enough for WSN applications, but they need a very accurate process characterization.

3.10 Conclusions

In order to investigate the feasibility of a time reference for WSNs, several possible physical principles have been analyzed in this chapter. As described in details in Chap. 2, such time reference must achieve an inaccuracy of the order of 1% while dissipating less than 100 μW. Moreover, it would be advantageous if the reference could be integrated at low costs using a digital CMOS process.

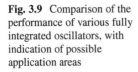
Fig. 3.9 Comparison of the performance of various fully integrated oscillators, with indication of possible application areas

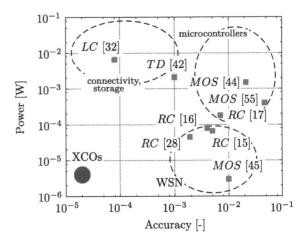

The required inaccuracy of 1% must be ensured regardless of all possible sources of error. In particular, it has been shown that a fully integrated time reference for WSNs is mainly affected by flicker noise and by PVT variations. Process variations can be compensated by a single-point trim at the expense of an increased cost. Robustness to supply voltage variations is usually addressed at the circuit-level design, since it is not strictly related to the reference principle as much as to its particular implementation. The temperature coefficient of the time reference can be effectively compensated if the intrinsic spread of the reference principle over temperature is enough low. Otherwise, multi-point trim is required, which is often too expensive.

Among the presented alternatives, TD-based frequency reference are sufficiently accurate even without any trim. However, together with LC-based reference, they are ruled out from the feasible choices by the excessive power consumption, which exceeds the 100-μW limit. MOS-based and RC-based references consume a much more WSN-compatible power. Nevertheless, the reliability and accuracy of MOS-based references strongly depends on the availability of accurate models for integrated devices, while the accuracy of RC-based reference relies on the availability of integrated resistors with low or accurately defined temperature coefficients. These reasons can limit their employment in a low-cost application.

Mobility-based time references may constitute a viable solution. They are based on a single physical phenomenon, i.e. charge mobility, and do not require any other accurate integrated component. No limit is foreseen in their power consumption with respect to RC or MOS references. The frequency accuracy which can be achieved by such references is investigated in the following chapter.

References

1. Jespersen J, Fitz-Randolph J (1977) From sundials to atomic clocks: understanding time and frequency. Dover Publications, Inc., New York
2. Lee TH (2008) It's about time: a brief chronology of chronometry. IEEE Solid State Circ, 13(3):42–49
3. IEEE (1999) IEEE standard definitions of physical quantities for fundamental frequency and time metrology – random instabilities, Std. 1139–1999
4. Razavi B (1998) RF microelectronics. Prentice-Hall, NJ
5. ITU-T (1997) Definitions and terminology for synchronization networks, Std. G810
6. Gierkink SL (1999) Control linearuty and jitter of relaxation oscillators. PhD thesis, University of Twente, http://icd.el.utwente.nl/research/index.php?id=12
7. Courant R, John F (1965) Introduction to calculus and analysis, vol 1. Wiley, NJ, pp 589–591
8. Demir A (2006) Computing timing jitter from phase noise spectra for oscillators and phase-locked loops with white and 1/f noise. IEEE Trans Circ Syst I 53(9):1869–1884. DOI 10.1109/TCSI.2006.881184
9. Liu C, McNeill J (2004) Jitter in oscillators with 1/f noise sources. Proc ISCAS 1:I–773–6. DOI 10.1109/ISCAS.2004.1328309
10. Gray PR, Hurst PJ, Lewis SH, Meyer RG (2001) Analysis and design of analog integrated circuits, 4th edn. Wiley, NJ
11. Hastings A (2005) The art of analog layout. Prentice Hall, NJ
12. Ueda N, Nishiyama E, Aota H, Watanabe H (2009) Evaluation of packaging-induced performance change for small-scale analog IC. IEEE Trans Semicond Manuf 22(1):103–109. DOI 10.1109/TSM.2008.2010739
13. Abesingha B, Rincon-Mora G, Briggs D (2002) Voltage shift in plastic-packaged bandgap references. IEEE Trans Circ Syst I 49(10):681–685. DOI 10.1109/TCSII.2002.806734
14. Ali H (1997) Stress-induced parametric shift in plastic packaged devices. IEEE Trans Comp Packag Manuf Technol B 20(4):458–462
15. De Smedt V, De Wit P, Vereecken W, Steyaert M (2009) A 66 μW 86 ppm/°C fully-integrated 6 MHz wienbridge oscillator with a 172 dB phase noise FOM. IEEE J Solid State Circ 44(7):1990–2001. DOI 10.1109/JSSC.2009.2021914
16. Lee J, Cho S (2009) A 10MHz 80μW 67 ppm/°C CMOS reference clock oscillator with a temperature compensated feedback loop in 0.18μm CMOS. In: 2009 Symposium on VLSI Circuits Dig. Tech. Papers, pp 226–227
17. Ueno K, Asai T, Amemiya Y (2009) A 10MHz 80μW 67 ppm/°C CMOS reference clock oscillator with a temperature compensated feedback loop in 0.18μm CMOS. In: Proc. ESSCIRC, pp 226–227
18. Pertijs MA, Huijsing JH (2006) Precision temperature sensors in CMOS technology. Springer, Dordrecht
19. Lane WA, TWrixon G (1989) The design of thin-film polysilicon resistors for analog IC applications. IEEE Trans Electron Dev 36(4):738–744. DOI 10.1109/16.22479
20. Lampard D (1957) A new theorem in electrostatics with applications to calculable standards of capacitance. Proc IEE C Monogr 104(6):271–280. DOI 10.1049/pi-c.1957.0032
21. Thompson A (1959) The cylindrical cross-capacitor as a calculable standard. Proc IEE B Electron Comm Eng 106(27):307–310. DOI 10.1049/pi-b-2.1959.0262
22. McCreary J (1981) Matching properties, and voltage and temperature dependence of mos capacitors. IEEE J Solid State Circ 16(6):608–616
23. St Onge S, Franz S, Puttlitz A, Kalinoski A, Johnson B, El-Kareh B (1992) Design of precision capacitors for analog applications. IEEE Trans Comp Hybrids Manuf Technol 15(6):1064–1071. DOI 10.1109/33.206932
24. Svelto F, Erratico P, Manzini S, Castello R (1999) A metal-oxide-semiconductor varactor. IEEE Electron Dev Lett 20(4):164–166. DOI 10.1109/55.753754

25. Chen KM, Huang GW, Wang SC, Yeh WK, Fang YK, Yang FL (2004) Characterization and modeling of SOI varactors at various temperatures. IEEE Trans Electron Dev 51(3):427–433. DOI 10.1109/TED.2003.822585

26. Sedra AS, Smith KC (1998) Microelectronics circuits, 4th edn. Oxford University Press, New York

27. Navid R, Lee T, Dutton R (2005) Minimum achievable phase noise of RC oscillators. IEEE J Solid State Circ 40(3):630–637. DOI 10.1109/JSSC.2005.843591

28. Tokunaga Y, Sakiyama S, Matsumoto A, Dosho S (2010) An on-chip CMOS relaxation oscillator with voltage averaging feedback. IEEE J Solid State Circ 45(6):1150–1158

29. McCorquodale MS, O'Day JD, Pernia SM, Carichner GA, Kubba S, Brown RB (2007) A monolithic and self-referenced RF LC clock generator compliant with USB 2.0. IEEE J Solid State Circ 42(2):385–399. DOI 10.1109/JSSC.2006.883337

30. McCorquodale MS, Pernia SM, O'Day JD, Carichner G, Marsman E, Nguyen N, Kubba S, Nguyen S, Kuhn J, Brown RB (2008) A 0.5-to-480 MHz self-referenced CMOS clock generator with 90 ppm total frequency error and spread-spectrum capability. In: ISSCC Dig. of Tech. Papers, pp 524–525

31. McCorquodale M, Carichner G, O'Day J, Pernia S, Kubba S, Marsman E, Kuhn J, Brown R (2009) A 25-MHz self-referenced solid-state frequency source suitable for XO-replacement. IEEE Trans Circ Syst I 56(5):943–956. DOI 10.1109/TCSI.2009.2016133

32. McCorquodale M, Gupta B, Armstrong W, Beaudouin R, Carichner G, Chaudhari P, Fayyaz N, Gaskin N, Kuhn J, Linebarger D, Marsman E, O'Day J, Pernia S, Senderowicz D (2010) A silicon die as a frequency source. In: IEEE International Frequency Control Symp., pp 103–108. DOI 10.1109/FREQ.2010.5556366

33. Groszkowski J (1964) Frequency of self-oscillations. Pergamon Press, Oxford

34. Groves R, Harame DL, Jadus D (1997) Temperature dependence of Q and inductance in spiral inductors fabricated in a silicon-germanium/BiCMOS technology. IEEE J Solid State Circ 32(9):1455–1459

35. Pouydebasque A, Charbuillet C, Gwoziecki R, Skotnicki T (2007) Refinement of the sub-threshold slope modelling for advanced bulk CMOS devices. IEEE Trans Electron Dev 54(10):2723–2729

36. Lee TH (2004) The design of CMOS Radio-frequency integrated circuits, 2nd edn. Cambridge University Press, Cambridge

37. Makinwa K, Snoeij M (2006) A CMOS temperature-to-frequency converter with inaccuracy of less than $0.5°C$ (3σ) from $-40°C$ to $105°C$. IEEE J Solid State Circ 41(12):2992–2997

38. Kashmiri SM, Pertijs MAP, Makinwa KAA (2010) A thermal-diffusivity-based frequency reference in standard CMOS with an absolute inaccuracy of $\pm0.1\%$ from $-55°C$ to $125°C$. IEEE J Solid State Circ 45(12):2510–2520

39. Makinwa KAA (2010) Smart temperature sensors in standard CMOS. In: Proc. Eurosensors XXIV, pp 930–939

40. van Vroonhoven C, d'Aquino D, Makinwa K (2010) A thermal-diffusivity-based temperature sensor with an untrimmed inaccuracy of $\pm0.2°C$ (3σ) from $-55°C$ to $125°C$. In: ISSCC Dig. of Tech. Papers, pp 314–315. DOI 10.1109/ISSCC.2010.5433900

41. Kashmiri S, Xia S, Makinwa K (2009) A temperature-to-digital converter based on an optimized electrothermal filter. IEEE J Solid State Circ 44(7):2026–2035. DOI 10.1109/JSSC.2009.2020248

42. Kashmiri SM, Souri K, Makinwa KAA (2011) A scaled thermal-diffusivity-based frequency reference in $0.16\mu m$ CMOS. In: Proc. ESSCIRC, pp 503–506

43. Kashmiri SM, Pertijs MAP, Makinwa KAA (2010) A thermal-diffusivity-based frequency reference in standard CMOS with an absolute inaccuracy of $\pm0.1\%$ from $-55°C$ to $125°C$. In: ISSCC Dig. of Tech. Papers, pp 74–75

44. Sundaresan K, Allen P, Ayazi F (2006) Process and temperature compensation in a 7-MHz CMOS clock oscillator. IEEE J Solid State Circ 41(2):433–442

45. Paavola M, Laiho M, Saukoski M, Halonen K (2006) A 3 μW, 2 MHz CMOS frequency reference for capacitive sensor applications. In: Proc. ISCAS, pp 4391–4394

46. Ge G, Zhang C, Hoogzaad G, Makinwa K (2010) A single-trim CMOS bandgap reference with a 3σ inaccuracy of $\pm0.15\%$ from $-40°$C to $125°$C. In: ISSCC Dig. of Tech. Papers, pp 78–79. DOI 10.1109/ISSCC.2010.5434040

47. Blauschild R (1994) An integrated time reference. ISSCC Dig. of Tech. Papers, pp 56–57

48. Jiang CL (1988) Temperature compensated monolithic delay circuit, US Patent 4843265

49. Tsividis Y (2003) Operation and modeling of the MOS transistor, 2nd edn. Oxford University Press, New York

50. Nguyen CC (2007) MEMS technology for timing and frequency control. IEEE Trans Ultrason Ferroelect Freq Contr 54(2):251–270. DOI 10.1109/TUFFC.2007.240

51. Ruffieux D, Krummenacher F, Pezous A, Spinola-Durante G (2010) Silicon resonator based 3.2 μW real time clock with 10 ppm frequency accuracy. IEEE J Solid State Circ 45(1):224–234. DOI 10.1109/JSSC.2009.2034434

52. Perrott M, Pamarti S, Hoffman E, Lee F, Mukherjee S, Lee C, Tsinker V, Perumal S, Soto B, Arumugam N, Garlepp B (2010) A low area, switched-resistor based fractional-n synthesizer applied to a MEMS-based programmable oscillator. IEEE J Solid State Circ 45(12):2566–2581. DOI 10.1109/JSSC.2010.2076570

53. SiTime Corporation (2009) SiT8003XT datasheet, Sunnyvale, CA. http://www.sitime.com. Accessed 23 Aug 2009

54. Discera Inc. (2009) DSC1018 datasheet, San Jose, CA. http://www.discera.com. Accessed 23 Aug 2009

55. Shyu YS, Wu JC (1999) A process and temperature compensated ring oscillator. In: Proc. Asia-Pacific Conference on ASICs, pp 283–286

Chapter 4
A Mobility-Based Time Reference

4.1 Introduction

As shown in Chap. 2, WSN nodes must be equipped with fully integrated time references with an accuracy of the order of 1% and a power consumption lower than $100\,\mu$W. Recently, much work has been devoted to implementing fully integrated time references in standard microelectronic technologies. As shown in Chap. 3, the inaccuracy of several of them is low enough for WSN applications, but they need either a too high power consumption or a very accurate process characterization, with a consequent limitation of their practical use.

As already pointed out at the end of Chap. 3, an alternative way of realizing an accurate fully integrated oscillator is by employing charge mobility as a reference. Mobility is less sensitive to process variations than other parameters, such as polysilicon resistance or oxide capacitance, and its standard deviation is less than 2% at room temperature for the process adopted for the experimental validation of this work. As noted in [1] and demonstrated by the data in [2], variations of the doping concentration only slightly affect the value of the mobility. Although the temperature dependence of the mobility is large (approximately $T^{-1.5}$), it is well defined and can be compensated for. The effect of process spread can then be removed by a single room-temperature trim.

Despite these desirable features of mobility-based references, no experimental data[1] about their accuracy has been reported in the open literature and, consequently, no point for mobility-based references is present in the design space of Fig. 3.9. To fill this gap, this chapter describes a fully integrated oscillator referenced to the electron mobility, which will be shown to be suitable for WSN applications. After a brief summary of the requirements, the principle of operation is described in Sect. 4.3. In the next section, a general analysis of the accuracy of the mobility-based

---

[1] A mobility-based time reference is presented in [1] but measurement results from only one sample are reported, which do not give any information about the spread.

F. Sebastiano et al., *Mobility-based Time References for Wireless Sensor Networks*,
Analog Circuits and Signal Processing, DOI 10.1007/978-1-4614-3483-2_4,
© Springer Science+Business Media New York 2013

reference is given. The practical circuit implementation, the experimental results and the temperature compensation strategies are presented in Sects. 4.5, 4.6 and 4.7, respectively. The effect of packaging and different process options are treated in Sect. 4.8.

4.2 Requirements

The requirements for time references for WSN have been broadly discussed in the previous chapters. Those requirements directly translate into the following specifications:

- *Accuracy*: The total inaccuracy over temperature and voltage variations and over multiple samples should be of the order of 1%.
- *Adopted IC process*: The reference should be realized in a deep-submicron technology (65-nm CMOS), to enable the integration of a complete WSN node, including RF and digital processing on a single chip (see Sect. 2.2).
- *Frequency of oscillation*: A frequency of 100 kHz has been chosen, which is low enough to achieve ultra-low power consumption and high enough to ensure sufficient time resolution.
- *Power consumption*: This must be lower than 50 μW.
- *Noise*: Thermal noise can be tolerated until it does not dominate the jitter performance in the time span of 1 s; flicker noise must be reduced until the relative jitter is below 0.2%, i.e. negligible with respect to the required accuracy of 1%.
- *Trimming*: A single-point trim is allowed to reduce the process spread.
- *Supply*: The minimum supply voltage must not exceed 1.2 V at a supply sensitivity of 1%/V.
- *Temperature coefficient*: The temperature coefficient of the reference must be dominated by the temperature coefficient of mobility and all others effects must be reduced accordingly.

Adopting those specifications, the accuracy of the reference will be dominated by the spread of the temperature coefficient of the mobility. Finding a bound for this spread is the aim of this chapter.

4.3 Principle of Operation

4.3.1 Deriving a Frequency from Carrier Mobility

As discussed in Sect. 3.7, a current proportional to the carrier mobility in a MOS transistor can be obtained by compensating for the transistor's threshold voltage, as

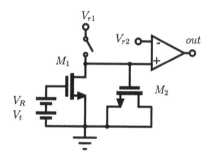

Fig. 4.1 Principle of
mobility-based time reference

schematically shown in Fig. 4.1. A mobility-dependent frequency can be obtained, for example, by periodically integrating such a current on a capacitor, as also shown by (3.44). This frequency is still dependent on the MOS oxide capacitance C_{ox} which is subject to temperature variation and process spread. In order to cancel this dependence, a MOS capacitor can be employed as the integrating capacitor. Using this simple principle, a mobility-based reference can be built. However, to ensure high accuracy, a proper topology must be chosen.

4.3.2 Oscillator Topology

Different oscillator topologies derived from the basic concept of charging and discharging a capacitor can be employed. However, since the aim is to find a bound for the accuracy of mobility-based oscillator, the simplest topology, which does not introduce excessive parasitic effects at a reasonably low power consumption, should be employed.

Adopting the classification for oscillator topologies already described in Sect. 3.3.2 for RC oscillators, three main categories can be identified: *linear*, *non-linear* and *locked* oscillators. A mobility-based linear oscillator can be realized by replacing the frequency-defining linear filter in Fig. 3.2, with a filter whose singularities depends on the mobility. This can be accomplished by using a $G_m - C$ filter and designing a circuit element whose transconductance is proportional to μ_n [3]. Such circuit element can be even a simple differential pair (whose transconductance is directly proportional to mobility) but care must be taken to appropriately compensate the threshold voltage (see Sects. 3.7 and 4.3.1). This can be achieved by biasing the circuit, for example the differential pair in the aforementioned example, with the proper voltage or current. The bias compensation voltage would be superimposed to the signal in the filter, resulting in a complex implementation. Instead, current biasing requires a current reference, which is also required in the simpler oscillators presented in the following. Moreover, as already stated in the previous chapter, the accuracy of linear oscillators can be limited by the required amplitude control loop. For those reasons, linear oscillators are not taken into further consideration.

Non-linear and locked oscillator are directly based on the circuit of Fig. 3.8, which can produce a pulse with duration inversely proportional to mobility. For non-linear oscillators, such as relaxation oscillators, this pulse is periodically generated to produce a continuous oscillation. In locked oscillators, the duration of the pulse is compared to the period of a second oscillator to tune its oscillation frequency. In both cases, the concept of Fig. 4.1 can be implemented in a straightforward way, by relying on the linear charge (or discharge) of capacitors with a constant current. By such simplification of the oscillator topology, the sources of inaccuracy are reduced to the minimum. In practice, the core of the oscillator would consist of two main blocks: a mobility-based current reference (represented in Fig. 4.1 by M_1 and the voltage sources V_t and V_R) and a capacitor-charging circuit (represented by MOS capacitor M_2, the comparator and the switches). Their accuracy is separately analyzed in the following sections. By inspection of the simple model in Fig. 4.1 and by assuming that one period of oscillation is equal to the time needed by the current through M_1 to discharge the MOS capacitor from V_{r1} to V_{r2}, the oscillation frequency can be written as:

$$f = \frac{I_1}{C_2(V_{r1} - V_{r2})} \tag{4.1}$$

where I_1 is the current through M_1 and proportional to electron mobility and C_2 is the gate capacitance of M_2. The proportionality of the frequency and the output current of the current reference, in this case represented by M_1, will be used in the error analysis.

4.4   Oscillator Accuracy

The various sources of inaccuracy of the time reference are analyzed in the following. Since it has been chosen to use the mobility as a reference, the accuracy must be limited by the spread of the mobility and the inaccuracy due to the other errors should be much less than the total target inaccuracy of 1%.

4.4.1   Trimming and Temperature Compensation

Trimming and temperature compensation will be described in details in this chapter in Sect. 4.7 and in Chap. 5. However, the basics are shortly introduced in the following, to identify which source of errors are dominant after trimming and temperature compensation.

Naming $f_0(T)$ the nominal uncompensated output frequency (as a function of the temperature) and $\Delta f(T)$ the frequency error due to process spread and mismatch, the output uncompensated frequency is given by

$$f(T) = f_0(T) + \Delta f(T) \tag{4.2}$$

The trimming of the individual sample at room temperature T_0 can be modeled as the division of the uncompensated frequency by the dimensionless factor $f(T_0)/f_0(T_0)$:

$$f_{trim}(T) = \frac{f(T)}{f(T_0)} \cdot f_0(T_0) \tag{4.3}$$

Regardless of their frequency error Δf, all samples have a trimmed output frequency at room temperature equal to $f_0(T_0)$. The temperature compensation of the trimmed frequency is modeled in a similar way, i.e. with the division by the factor $f_0(T)/f_0(T_0)$:

$$f_{comp}(T) = \frac{f_{trim}(T)}{f_0(T)} \cdot f_0(T_0) = \frac{f(T)}{f_0(T)f(T_0)} \cdot [f_0(T_0)]^2 \tag{4.4}$$

The effects of compensation non-idealities, such as a limited trimming resolution, will be treated in the aforementioned sections.

The relative error in the compensated frequency can be derived using (4.2) and (4.4), as

$$\frac{\Delta f_{comp}(T)}{f_0(T_0)} = \frac{f_{comp}(T) - f_0(T_0)}{f_0(T_0)} \tag{4.5}$$

$$= \frac{1 + \frac{\Delta f(T)}{f_0(T)}}{1 + \frac{\Delta f(T_0)}{f_0(T_0)}} - 1 = \frac{\frac{\Delta f(T)}{f_0(T)} - \frac{\Delta f(T_0)}{f_0(T_0)}}{1 + \frac{\Delta f(T_0)}{f_0(T_0)}} \tag{4.6}$$

$$\approx \frac{\Delta f(T)}{f_0(T)} - \frac{\Delta f(T_0)}{f_0(T_0)} \tag{4.7}$$

where it has been assumed that $\frac{\Delta f(T_0)}{f_0(T_0)} \ll 1$. Equation (4.7) can be interpreted as the variation of the relative frequency error $\frac{\Delta f}{f_0}$ over temperature. In practice, even if the frequency error Δf is non-zero, the error on the *trimmed* and *temperature-compensated* frequency can be zero if the temperature coefficient of $\frac{\Delta f(T)}{f_0(T)}$ is zero. This will have an impact on the analysis of the error sources and will imply that for some error sources it is more important to have a very low temperature coefficient than a low absolute value.

4.4.2 Topology of the Current Reference

To find the minimum number of circuit elements required for the implementation of the mobility-based current reference, it is necessary to start from the input-output characteristic of a MOS transistor, shown in Fig. 4.2 for a MOS in saturation.

Fig. 4.2 Input-output
characteristic of a MOS
transistor

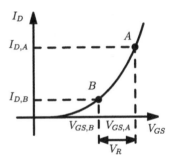

As an approximation, we assume that the current can be expressed as

$$I_D = \mu_n g(V_{GS} - V_t) \tag{4.8}$$

where I_D is the drain current, V_{GS} is the gate-source voltage, V_t is the threshold
voltage and $g(\cdot)$ is a function whose exact shape is not significant here. In the
simple square-law MOS model, it is approximated with a second-order polynomial,
i.e. $g(V) = \frac{1}{2}C_{ox}\frac{W}{L}V^2$. Thus, two main assumptions are implicit in (4.8): first,
the current is proportional to the mobility; second, a change in threshold voltage
causes a shift of the curve along the x-axis. Due to the latter effect, determining a
single point on the curve, such as point A in Fig. 4.2, by biasing the transistor with a
reference voltage $V_{GS,A}$, is not sufficient to make the current $I_{D,A}$ independent from
variations of the threshold voltage. Instead, two points, A and B, are needed, which
can be defined on the $I_D - V_{GS}$ curve by two independent conditions, one on their
x-coordinates (V_{GS}) and one on their y-coordinates (I_D). As a first condition, it is
practical to use the following

$$V_{GS,A} - V_{GS,B} = V_R \tag{4.9}$$

since it is possible to accurately generate an absolute reference voltage on a silicon
chip (see Sect. 3.7). For the conditions on the currents, it is advantageous to use

$$\frac{I_{D,A}}{I_{D,B}} = n \tag{4.10}$$

thanks to the possibility of realizing in microelectronic technologies very accurately
matched components and, consequently, very accurate ratios. For practical MOS
characteristics and parameters, conditions (4.9) and (4.10) ensure the uniqueness of
the couple of points A and B. Thus, both $I_{D,A}$ and $I_{D,B}$ are currents proportional to
mobility and independent of the threshold voltage [4].

To make the previous statement true, the assumptions made in using (4.8) must
be ensured. In particular, the operation of the MOS transistor must be described with
sufficient accuracy by (4.8). The mobility is dependent on the region of operation of
the device, and in particular on the applied gate-source voltage. As a consequence,

Fig. 4.3 Principle of operation of the current reference

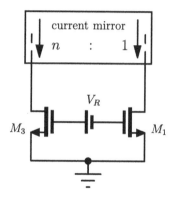

the reference voltage V_R must be kept low enough to ensure that the mobility is the same both for point A and B. In different regions of operation, the mobility is determined by different physical phenomenons, among which phonon scattering, Coulomb scattering and surface-roughness scattering, which show different temperature dependence [5]. In order to get a unique phenomenon dominating the temperature dependence of mobility and, consequently, of the oscillator frequency, V_R must be limited to few hundreds of mV. Devices from the 65-nm CMOS process described in Sect. 4.6 have been simulated and results show that for V_R between 0.15 and 0.2 V, the 3σ-inaccuracy due to this effect is less than 0.8% over the military temperature range.

In addition, the accuracy of the current generated with the above procedure can be increased if the function $g(\cdot)$ shows low dependence with respect to other parameters. This is why it is chosen to bias the transistor in strong inversion and in saturation: respectively, to lower as much as possible the influence of the threshold voltage (which has an exponential dependence on the drain voltage in weak inversion) and to lower the influence of drain-source voltage.

Regarding the implementation of the circuit which generates the reference current based on the previous strategy, it could be possible to bias a single device alternatively at point A and B with the proper voltages and currents and storing in the analog domain the bias point. However, this could cause switching transients and switching errors, such as charge-injection-related errors, which can limit the performance. Thus, it is better to use two devices (not necessarily equal), one biased at point A or the other at point B. The difference in V_{GS} can be generated by short-circuiting the gates and implementing a voltage source V_R between the transistors' sources. However, such voltage source should be able to generate an accurate voltage while absorbing the drain current of at least one of the two transistors. It is then chosen to short-circuit the sources and use a voltage source in series with the gates, as shown in Fig. 4.3, since gate currents are very small even in deep-submicron technologies. The second condition (4.10) is enforced by the current mirror shown in the figure. Note that the circuit in Fig. 4.3 is only used to show the principle of operation. A complete schematic which works in practice will be presented in Sect. 4.5.

To simplify the analysis of the errors introduced by the $M_1 - M_2$ pair and the current mirror, the square-law MOS model is used. It is possible to derive [4]

$$I_1 = \frac{\mu_n C_{ox}}{2} \frac{W_1}{L_1} \frac{V_R^2}{\left(\sqrt{\frac{\pi}{m}} - 1\right)^2} \tag{4.11}$$

where μ_n is the electron mobility, C_{ox} is the gate capacitance per unit area and $m = \frac{W_3/L_3}{W_1/L_1}$. In the previous equation and in the rest of the chapter, the symbols I_x, W_x and L_x are used for the drain current, width and length, respectively, of transistor M_x.

As proven in Appendix A, the relative error in the current is limited by

$$\frac{\Delta I_1}{I_1} \leq 2 \frac{\Delta V_{t1,3}}{V_R} + \left(\frac{n}{m} - \sqrt{\frac{n}{m}}\right)\left[\frac{\Delta n}{n} + \frac{\Delta \beta_{1,3}}{\beta_{1,3}} + \max\{\lambda_1, \lambda_3\}(V_{DS1} - V_{DS3})\right] \tag{4.12}$$

where $\Delta V_{t1,3}$ is the mismatch between the threshold voltages of M_1 and M_3, $\frac{\Delta \beta_{1,3}}{\beta_{1,3}}$ is the mismatch between the β factor of M_1 and M_3, $\frac{\Delta n}{n}$ is the error in the current mirror gain and $\lambda_{1,3}$ are the channel-length-modulation parameters for $M_{1,3}$ (for exact definitions the reader is referred to the appendix). For reasons concerning the stability of the circuit (see Sect. 4.5) and, consequently, its practical implementation, $\frac{n}{m} > 1$. However, to increase the accuracy, it is clear that $\frac{n}{m}$ must be minimized. As a trade-off and to enable an estimation of the order of magnitude of the different errors, $\frac{n}{m}$ is considered to be smaller than five. In the following, the different sources of error in (4.12) are individually analyzed.

4.4.3 Accuracy of the Current Reference

Mismatch of M_1 and M_3

By combining the two equations (4.7) and (4.12) and by considering only the matching errors of M_1 and M_3, the inaccuracy of the frequency after compensation can be written as

$$\frac{f_{comp}(T) - f_0(T_0)}{f_0(T_0)} = \frac{\Delta I_1(T)}{I_{1,0}(T)} - \frac{\Delta I_1(T_0)}{I_{1,0}(T_0)} = 2 \frac{\Delta V_{t1,2}(T) - \Delta V_{t1,2}(T_0)}{V_R} +$$
$$+ \left(\frac{n}{m} - \sqrt{\frac{n}{m}}\right)\left(\frac{\Delta \beta_{1,3}(T)}{\beta_{1,3}(T)} - \frac{\Delta \beta_{1,3}(T_0)}{\beta_{1,3}(T_0)}\right) \tag{4.13}$$

As discussed in the appendix in Sect. A.4, data for the temperature behavior of matching parameters are usually unavailable and the following discussion is based on approximate figures derived in the appendix. Based on those consideration, it is assumed that $\Delta V_{t1,2}(T) - \Delta V_{t1,2}(T_0) < 100\,\mu V$ and $\frac{\Delta\beta_{1,3}(T)}{\beta_{1,3}(T)} - \frac{\Delta\beta_{1,3}(T_0)}{\beta_{1,3}(T_0)} < 0.2\%$. Since both the current factor mismatch and the threshold voltage mismatch are stochastic phenomenon, their independence is assumed in the computation of their cumulative contribution. Thus, taking into account the bounds on the value of $\frac{n}{m}$ previously described and the fact that practically $150\,mV < V_R < 300\,mV$ (to keep $M_{1,3}$ in strong inversion and in the same operation region), the frequency inaccuracy will be lower than approximately 0.5%.

Leakage

Leakage in a CMOS circuit is mainly due to the currents through reverse-biased junctions, such as drain and source junctions, to gate leakage and to subthreshold leakage. Subthreshold leakage currents are present in MOS transistors with very low V_{GS}, which are intended to be inactive by design. Subthreshold leakage can be minimized by reducing, or even avoiding, the number of non-active transistors connected by their drain or source to the sensitive nodes of the circuit. As proven in the circuit implementation described in this chapter, this is possible for the mobility-based current reference, so that subthreshold leakage can be neglected.

Gate leakage, i.e. the tunneling current through the gate oxide, is usually not accurately modeled and it is difficult to quantify its impact on the performance. However, employing only transistors with thick gate oxide, reduce the gate leakage. For example, for gate oxide thicknesses greater than 2 nm, typical of technologies with feature size larger than $0.18\,\mu m$ [6], the leakage is less than about $0.1\,nA/\mu m^2$ for voltage differences across the oxide less than 1 V [7,8], hence resulting in errors lower than 0.1% for bias currents of the order of 100 nA. Thick oxide transistors are usually available in the baseline version of deep-submicron processes to allow off-chip interfacing with different voltage domains. For example, the 65-nm technology described in the experimental section of this chapter offers the availability of 2.5-V thick oxide transistors, for which gate leakage is well below $1\,fA/\mu m^2$. Thus, by proper use of thick oxide device, gate leakage can be neglected.

On the contrary, junction leakage can not be avoided. Even if some estimates are in the order of a few fA [9], leakage currents per drain width below 100°C can be up to $1\,nA/\mu m$ [10]. In Appendix A, it has been shown that the error due to leakage currents I_{l1} and I_{l3} at the drains of M_1 and M_3, respectively, is

$$\frac{\Delta I_1}{I_1} \approx \left(\frac{n}{m} - \sqrt{\frac{n}{m}}\right)\frac{nI_{l1} - I_{l3}}{nI_1} \tag{4.14}$$

Thus, it is sufficient to keep the currents of interest, such as I_1 and I_3, larger than $1\,\mu A$ to ensure an inaccuracy of less than 0.2% for a maximum expected leakage of $1\,nA/\mu m$. In the adopted process, much lower leakage currents are expected, resulting in the choice of the minimum current of the order of 100 nA.

Fig. 4.4 Basic current
reference with current mirror

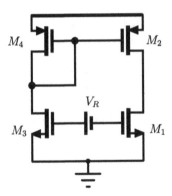

Finite Output Impedance

If the drains of M_1 and M_3 are kept at a different potential, their current is affected
by their limited output impedance $r_0 = \lambda^{-1} I_D^{-1}$. The channel-length-modulation
parameter λ is approximately inversely proportional to the length of the transistor.
For a fixed drain current and keeping the transistor in saturation, the length can be
increased by scaling L and W proportionally, i.e. by increasing the device area. For
a device area below $100\,\mu m^2$, it is possible to keep L in the order of a few tens of
μm and λ is in the order of $0.01\,V^{-1}$ for a typical CMOS process [11].

Furthermore, the voltages at the drains of M_1 and M_3 can be kept equal by an
additional opamp (as shown in the circuit implementation). Their difference will
then be approximately equal to the opamp's input offset voltage, which can be easily
limited to less than $10\,$mV. The current error due to the finite output impedance is
then less than 0.03%.

Errors in the Current Mirror

In the following we assume that a simple pMOS current mirror is employed to
implement the current ratio n, as in Fig. 4.4. In the appendix, it is shown that the
error in the factor n can be expressed as

$$\frac{\Delta n}{n} \approx \frac{\Delta \beta_{2,4}}{\beta_{2,4}} + 2\frac{\Delta V_{t2,4}}{V_{GSp} - V_{tp}} \tag{4.15}$$

where all parameters refer to the pMOS pair used in the mirror. The effect of β
mismatch is the same as the β mismatch of $M_{1,3}$ and it is not treated again. Using
(4.7), the frequency error after trimming and due only to current mirror errors can
be written as

$$\frac{f_{comp}(T) - f_0(T_0)}{f_0(T_0)} = \frac{\Delta f(T)}{f_0(T)} - \frac{\Delta f(T_0)}{f_0(T_0)} \tag{4.16}$$

$$= \left(\frac{n}{m} - \sqrt{\frac{n}{m}}\right)\left(\frac{\Delta n(T)}{n(T)} - \frac{\Delta n(T_0)}{n(T_0)}\right) \tag{4.17}$$

$$= 2\left(\frac{n}{m} - \sqrt{\frac{n}{m}}\right)\left(\frac{\Delta V_{t2,4}(T)}{V_{GSp}(T) - V_{tp}(T)} - \frac{\Delta V_{t2,4}(T_0)}{V_{GSp}(T_0) - V_{tp}(T_0)}\right) \tag{4.18}$$

In the previous equation, the error due to finite output resistance of the transistor is neglected, following the result of the previous section. Moreover, such effect can be made even smaller by employing cascoding. If the mirror transistors are in strong inversion and using (4.11), we have

$$V_{GSp} - V_{tp} = \sqrt{\frac{2I_1(T)}{\beta_2(T)}} = V_{mirr}\sqrt{\frac{\mu_n(T)}{\mu_p(T)}} \tag{4.19}$$

where $V_{mirr} = \sqrt{\frac{W_1}{W_2}\frac{L_2}{L_1}}V_R\left(\sqrt{\frac{\pi}{m}} - 1\right)^{-1}$ is constant over temperature assuming that V_R is a temperature-independent reference voltage. By adopting the commonly used models $\mu_n = \mu_{n0}\left(\frac{T}{T_{ref}}\right)^{\alpha_\mu}$ and $\mu_p = \mu_{p0}\left(\frac{T}{T_{ref}}\right)^{\alpha'_\mu}$, the frequency error then becomes

$$\frac{f_{comp}(T) - f_0(T_0)}{f_0(T_0)} = 2\left(\frac{n}{m} - \sqrt{\frac{n}{m}}\right)\frac{\Delta V_{t2,4}(T_0)}{V_{GSp}(T_0) - V_{pt}(T_0)} \cdot$$

$$\cdot\left[\frac{\Delta V_{t2,4}(T)}{\Delta V_{t2,4}(T_0)}\cdot\left(\frac{T}{T_0}\right)^{\frac{\alpha'_\mu - \alpha_\mu}{2}} - 1\right] \tag{4.20}$$

Considering that values for α_μ and α'_μ commonly vary from -1.2 to -2 [12], the effect of the temperature behavior of the mobility will only worsen the temperature variation of the threshold mismatch by about 20% over the military temperature range. Thus, considering the typical value of the different parameters in (4.20), the effect of the mismatch of the current mirror is in the same order of the effect of the mismatch of M_1 and M_2.

4.4.4 Accuracy of the Oscillator Core

After generating the mobility-based current with sufficient accuracy, the current must be copied from the current reference, integrated in a capacitor and the threshold-crossing of the capacitor must be detected. The accuracy of each of these steps is analyzed in the following.

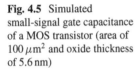

Fig. 4.5 Simulated small-signal gate capacitance of a MOS transistor (area of $100\,\mu m^2$ and oxide thickness of 5.6 nm)

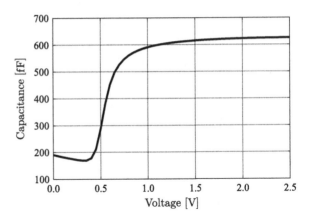

Threshold Detection

The error introduced by the threshold detection can be either an amplitude error, such as an offset in the comparator used for the detection, or a timing error, such as a delay between the effective crossing of the threshold and the operation of the oscillator logic circuitry. The effect of the latter can be minimized by using a sufficiently low frequency. For an oscillation frequency in the order of 100 kHz, considering two crossings per period (as explained in the implementation section), it is sufficient to ensure a delay smaller than 10 ns to keep the error below 0.2%. Note that the use of a low oscillation frequency relaxes the requirements on the speed of the comparator and makes it possible a low power implementation.

The amplitude error is mostly due to the offset of the comparator. For a given offset voltage, the inaccuracy has a lower bound determined by the voltage swing across the oscillator capacitor and, consequently, by the available supply voltage. For a voltage swing of about 0.5 V, the offset must be lower than $500\,\mu V$ for a frequency error of 0.1%.

Capacitor

As schematically shown in Fig. 4.1, a MOS capacitor is used as the integration capacitor of the mobility-based relaxation oscillator. Since the output of the current reference is proportional to C_{ox} [see (4.11)], the output frequency would be independent of C_{ox} if the capacitance of the MOS capacitor M_2 in Fig. 4.1 is proportional to C_{ox}. The *small-signal* gate capacitance of a MOS transistor is often approximated as $C = W \cdot L \cdot C_{ox}$, but this is valid only for devices biased in strong inversion (or in accumulation), i.e. for large enough $V_{GS} - V_t$, as shown in Fig. 4.5. Since the low-voltage requirements for the reference strongly limit the use of large overdrive voltages on the capacitor, the aforementioned approximation is not accurate enough.

It is well-known [12] that for a transistor in moderate or strong inversion, the lower the voltage on the gate, the more dependent C becomes on the threshold voltage. To reduce this source of error, the voltage across the capacitor, which is always comprised between V_{r1} and V_{r2}, should be as high as possible. Thus, the highest possible voltage is chosen for V_{r1}, i.e. $V_{r1} = V_{dd}$. Since a larger voltage swing on the capacitor reduces the effect of other error sources, such as the comparator offset, and a low V_{r2} affect the value of C, a trade-off exist in the choice if V_{r2}. An analytical treatment of the effect of the voltage range on the capacitor on the accuracy is shown in Appendix A, together with results from simulations of the 65-nm CMOS implementation. By adopting a voltage swing between 0.8 and 1.2 V, the inaccuracy introduced by a non-ideal MOS capacitor is in the order of 0.1%.

Current Mirrors

The current charging (or discharging) the capacitor must be copied from the current reference core of Fig. 4.4. This can be accomplished by a simple current mirror composed by transistor M_1 of the current reference (Fig. 4.3) and a second transistor M_A connected to the oscillator capacitor. Alternatively, a pMOS mirror can be built by using a transistor matched to M_2 and M_4. In both cases, since the frequency is proportional to the current mirror gain, the error in the trimmed and temperature-compensated frequency $f_{comp}(T)$ is simply given by the temperature variations of the mismatch of such current mirror. Such error can be computed in the same way as done for the $M_{2,4}$ mirror and for the matched pair $M_{1,3}$. Following the same reasoning of Sect. 4.4.2, the error can be considered in the same order of the errors in the current reference.

4.4.5 Residual Source of Error

If the previously described design techniques are adopted, the spread of the output frequency is only dominated by the spread of mobility. Charge mobility is usually modeled as

$$\mu_n(T) = \mu_0 \left(\frac{T}{T_0} \right)^{\alpha_\mu} \tag{4.21}$$

where μ_0 is the mobility at room temperature T_0 and α_μ typically varies from -1.2 to -2 [12]. Process variations can in general cause the spread of two parameters: the room-temperature mobility μ_0 and the temperature exponent α_μ. Assuming the output frequency proportional to the mobility, the uncompensated and untrimmed frequency is given by

$$f(T) = k \cdot (\mu_0 + \Delta\mu_0) \left(\frac{T}{T_0} \right)^{\alpha_\mu + \Delta\alpha} \tag{4.22}$$

where k is a proportionality constant. The nominal output frequency is simply

$$f_0(T) = k \cdot \mu_0 \left(\frac{T}{T_0}\right)^{\alpha_\mu} \tag{4.23}$$

An expression for the temperature-compensated trimmed output frequency can be derived by using (4.4), (4.22) and (4.23):

$$f_{comp}(T) = k \cdot \mu_0 \left(\frac{T}{T_0}\right)^{\Delta\alpha} \tag{4.24}$$

Note that the effect of the spread of μ_0 has been completely compensated by the trimming. Consequently, under the assumption that the spread of the mobility dominates the frequency accuracy, the spread in the temperature exponent α_μ determines the spread of the frequency error. In particular, it is possible to calculate the relative error of the compensated frequency:

$$\frac{\Delta f_{comp}(T)}{f_0(T_0)} = \frac{f_{comp}(T) - f_0(T_0)}{f_0(T_0)} \approx \Delta\alpha \frac{1}{f_0(T_0)} \left.\frac{\partial \Delta f_{comp}}{\partial \Delta\alpha}\right|_{\Delta\alpha=0} = \Delta\alpha \log\left(\frac{T}{T_0}\right) \tag{4.25}$$

4.5 Circuit Description

4.5.1 Oscillator Structure

A simplified schematic of the oscillator is shown in Fig. 4.6 [13]. It consists of a current reference, two current mirrors $M_1 - M_A$ and $M_1 - M_B$, two capacitors C_A and C_B and a comparator. The drain current of M_1 is mirrored by M_A and M_B with a gain of four and, as explained in the next section, is given by

$$I_1 = \frac{I_A}{4} = \frac{I_B}{4} = \frac{\mu_n C_{ox}}{2} \frac{W_1}{L_1} k V_R^2 \tag{4.26}$$

where μ_n is the electron mobility, C_{ox} is the gate capacitance per unit area, k is a constant determined by the ratios of the dimensions of matched transistors and V_R is a reference voltage. As shown in the timing diagram in Fig. 4.7, C_A and C_B are alternatively precharged to V_{r1} and then linearly discharged by M_A and M_B. When the voltage on the discharging capacitor drops below V_{r2}, the output of the comparator switches and the linear discharge of the other capacitor starts immediately, while the recharge is delayed by D. The delay D ensures that non-idealities of the comparator do not affect the slope of the discharge at the crossing

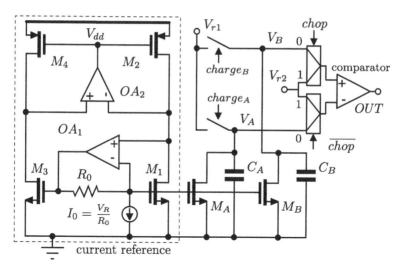

Fig. 4.6 Mobility-referenced oscillator

Fig. 4.7 Oscillator
waveforms

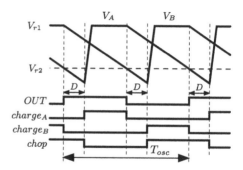

of V_{r2} and it is not critical, as it does not influence the period T_{osc}. The delay D and the signals driving the switches are generated by a digital circuit not shown in Fig. 4.6.

By inspecting Figs. 4.6 and 4.7, the period and frequency of oscillation can be easily determined, and from (4.26):

$$T_{osc} = \frac{2C}{I_A}(V_{r1} - V_{r2}) \Rightarrow f = \frac{1}{T_{osc}} = \mu_n k \frac{C_{ox}\frac{W_1}{L_1}}{C} \frac{V_R^2}{V_{r1} - V_{r2}} \qquad (4.27)$$

where $C = C_A = C_B$. C_A and C_B are MOS capacitors operating in inversion, in order to obtain a ratio $\frac{C_{ox}\frac{W_1}{L_1}}{C}$ which is independent of the effect of temperature and process variations on C_{ox}. If the reference voltages V_{r1} and V_{r2} are obtained from a bandgap reference, the residual frequency variations will be due to the spread[2]

---

[2]Geometric factors in (4.27) (W_1, L_1 and the area of capacitors C_A and C_B) are also affected by process spread. However, their effect on T_{osc} can be neglected if sufficiently large devices are employed.

and temperature dependence of the mobility and of the voltage V_R. The latter can be used as a control voltage to compensate for the effects of temperature variations and process spread, or it can be derived from a voltage reference like V_{r1} and V_{r2}. Further details on the use of V_R to compensate for temperature variations are given in Sect. 4.7.

The two multiplexers at the input of the comparator are driven by the signal *chop*, shown in Fig. 4.7, to mitigate the effect of comparator offset. Thus, with an offset V_{os} at the comparator input, the output is switched when $V_A = V_{r2} - V_{os}$ or $V_B = V_{r2} + V_{os}$ and the total error in the period is given by

$$\frac{\Delta t}{T_{osc}} \cong \frac{V_{os}}{2(V_{r1} - V_{r2})} \left(\frac{\Delta C}{C} - \frac{\Delta I}{I_A} \right) \tag{4.28}$$

where $\Delta C = C_A - C_B$ and $\Delta I = I_A - I_B$. Hence, if the capacitors and current mirrors are well matched, the resulting error is small. Since the oscillator is intended to operate at relatively low frequencies, good matching can be obtained by increasing device area without significantly affecting oscillator's performance.

4.5.2   Current Reference

The operation of the circuit in the dashed box in Fig. 4.6 can be understood by noting that M_2, M_4 and OA_2 constitute a low-voltage current mirror and that M_1 is effectively diode-connected through OA_1 and R. Using the square-law MOS model, it is possible to derive [4]

$$I_1 = \frac{\mu_n C_{ox}}{2} \frac{W_1}{L_1} \frac{V_R^2}{\left(\sqrt{\frac{n}{m}} - 1 \right)^2} \tag{4.29}$$

where $n = \frac{W_4/L_4}{W_2/L_2}$ and $m = \frac{W_3/L_3}{W_1/L_1}$ and, with reference to (4.26), $k = \left(\sqrt{\frac{n}{m}} - 1 \right)^{-2}$. Equation (4.11) is valid under the assumption that M_1 and M_3 are biased in strong inversion. It is also preferable to bias M_2 and M_4 in strong inversion in order to reduce the error due to mismatch in the current mirror gain n. The addition of OA_1 and OA_2 to the basic structure constituted by $M_1 - M_4$ increases the power consumption but at the same time reduces the requirement on voltage headroom (avoiding for example the diode-connection of M_4) and consequently reduces the minimum required voltage supply. Moreover, OA_2 increases the output impedance seen at the drain of M_4, which mitigates the effect of voltage supply variation on the scurrent reference output and hence the output frequency.

The complete schematic of the current reference is shown in Fig. 4.8. The current source I_0 is implemented by the unity-gain cascode current mirror $M_5 - M_8$. The value of I_0 is fixed by the current mirror $M_9 - M_{12}$ and by the external opamp,[3] which forces a voltage drop V_R on R_1, so that $R_0 I_0 = \frac{R_0}{R_1} \frac{W_9/L_9}{W_{11}/L_{11}} \frac{W_7/L_7}{W_5/L_5} V_R = V_R$. Resistance values ($R_0 = 200\,\mathrm{k}\Omega$, $R_1 = 20\,\mathrm{k}\Omega$) are chosen as a tradeoff between resistor area, current consumption and the contribution of the parasitic currents through R_1.[4]

The start-up circuit and the implementation of the opamps are shown in the dashed boxes in the figure. A folded cascode structure is adopted for OA_2 to reduce its systematic input offset. Since OA_1 must provide an output quiescent current I_0, it is biased with $I_{17} = I_0/2$ and it is dimensioned such that $\frac{W_{13}}{L_{13}} = \frac{W_{14}}{L_{14}}$ and $5\frac{W_{15}}{L_{15}} = \frac{W_{16}}{L_{16}}$. Both input pairs, $M_{13} - M_{14}$ and $M_{18} - M_{19}$, are biased in weak inversion to allow the input common mode of the opamps to be equal to V_{GS1}.

The stability of the whole circuit can be guaranteed by ensuring the stability of the two negative feedback loops, i.e. the one constituted by OA_1 and M_1 and the one constituted by OA_2 and M_4, and of the positive feedback loop formed by M_1, M_2 and the low-voltage current mirror. The first negative feedback loop can be modelled as the cascade of OA_1 in buffer configuration through R_0 and the common-source amplifier M_1. The bandwidth of OA_1 is much larger than the frequency of the pole associated to the output impedance of M_1, since OA_1 is biased with a larger current (with reference to Fig. 4.6, $I_0 = 1\,\mu\mathrm{A}$ and $I_1 = 125\,\mathrm{nA}$ in the nominal case as will be shown in Sect. 4.6) and MOS capacitor M_{27} is added at the drain of M_1. Miller compensation is used for the second loop, employing capacitor C_c across M_4. A fringe metal capacitor is needed to implement C_c because the voltage headroom available between gate and drain of M_4 is not enough to bias a MOS capacitor. Assuming that the DC open-loop gain of the two negative loops is high enough, the open-loop gain of the positive feedback loop $A_{loop}(f)$ at DC can be approximated as

$$A_{loop}(0) \approx \frac{1}{n}\frac{g_{m3}}{g_{m1}} = \frac{1}{n}\sqrt{\frac{W_3/L_3}{W_1/L_1}}\sqrt{\frac{I_3}{I_1}} = \sqrt{\frac{m}{n}} \tag{4.30}$$

where g_{m1} and g_{m3} are respectively the transconductance of M_1 and M_3 in Fig. 4.8. The loop is stable under the condition that $A_{loop}(0) < 1$ and that $A_{loop}(f)$ is monotonically decreasing, i.e. no peaking occurs in the frequency response of the loop. Since the negative feedback loops interact with the $M_1 - M_4$ loop, their phase

---

[3] An external opamp (LTC1053) is used only for testing purpose.

[4] Note that a pad with large ESD protection diodes and an external opamp are connected to one end of R_1. The parasitic current through R_1 is the sum of the leakage currents of the ESD diodes and of the input bias current of the opamp.

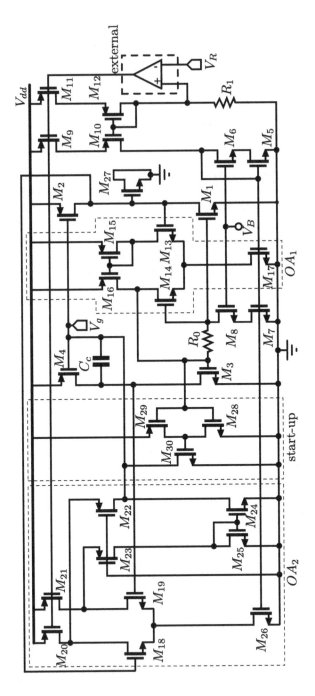

Fig. 4.8 Complete schematic of the current reference

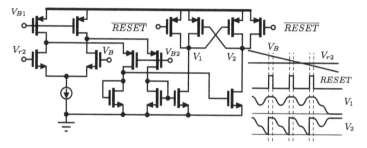

Fig. 4.9 Simplified schematic of autolatch comparator for $chop = 0$

margin should be large enough not only to ensure stability of the respective loops but also to prevent such peaking. In the presented circuit, the phase margins of the two loops were designed to be larger than 45° in the worst case (process corners and temperature).

To avoid coupling digital noise to the gate of M_1 via the output mirrors $M_1 - M_{A,B}$ of Fig. 4.6, the current is mirrored to M_A and M_B using the node V_G and additional pMOS and nMOS mirrors (not shown in Fig. 4.8).

4.5.3 Comparator

The delay of the comparator must be negligible with respect to the oscillation period T. This requires high gain and large bandwidth in the case of an open-loop topology, or a very small hysteresis in the case of a Schmitt trigger implementation. To overcome this problem, within the constraints of a very tight power budget, a comparator with novel architecture, called an *autolatch comparator*, was introduced. Its schematic is shown in Fig. 4.9 for the case when $chop = 0$ together with some waveforms. The core of the circuit is a dynamic latch. When a comparison is needed, a digital circuit resets the latch and then enables it. As long as V_B has not crossed V_{r2}, V_1 goes periodically to V_{dd} and V_2 to ground. The signal on V_2 is inverted and delayed through a chain of inverter gates to generate the $RESET$ signal. V_1 and V_2 are then pulled up to V_{dd} and, after a delay, $RESET$ goes low. This cycle is repeated until V_B crosses V_{r2} and V_1 goes low. In this case the output is represented by the voltage on V_1. When $chop = 1$, the logic takes care of generating $RESET$ from the appropriate node and chooses the right output node. The latch is preceded by a folded preamplifier to prevent kickback noise appearing on oscillator's capacitors.

The delay of the comparator can be adjusted by controlling the period of the described cycle. Simulations show that the delay is less than 13 ns in the worst case (process and temperature) with a total average current of 30 μA at 1.2 V supply. Low power is achieved by keeping the devices small, so as to minimize their parasitic capacitance. Small devices have high flicker noise, but the offset compensation technique (chopping) described in Sect. 4.5.1 also reduces the effect of flicker noise.

4.6 Experimental Results

The oscillator has been realized in a baseline TSMC 65-nm CMOS process [13]. The circuit occupies $0.11\,mm^2$ and uses only 2.5-V I/O thick-oxide MOS devices. 1.2-V thin oxide devices were avoided because of their high gate leakage current, which is significant in this circuit and which represents a significant fraction of the drain current for very long devices [14]. Most of the area of the circuit is occupied by the current reference and by the oscillator's capacitors[5] (Fig. 4.10). The layout of the current reference has been optimized for transistor matching; particular care has been devoted to reduce systematic mismatch due to topography related errors [15] and to reduce mechanical strain associated with metal chemical mechanical polishing (CMP) dummy structures [16,17]. To deal with the first effect, asymmetries in the surroundings of matched transistor-arrays are located more than $10\,\mu m$ away from the active devices. This precaution significantly increased the area of the current reference and was adopted to minimize additional sources of inaccuracy in this test chip; a substantial fraction of this area can probably be saved in a future redesign. In cases where dummy metal structures *had* to be included above arrays of matched transistors (to satisfy metal-density rules), such structures were manually drawn to minimize the mismatch due to the additional stress. For flexibility in testing, all the reference voltages (V_{r1}, V_{r2} and V_R) were provided externally. For a nominal oscillation frequency of approximately 100 kHz, the reference current is $I_1 =125\,nA$ for $C \cong 6\,pF$, $V_R =0.2\,V$, $V_{ref1} =1\,V$ and $V_{ref2} =0.6\,V$. A low frequency was chosen to reduce the impact of parasitic effects, such as comparator delay. The total current consumption with 1.2 V supply voltage is $34.3\,\mu A$ ($18.9\,\mu A$ for comparator and logic; $14.4\,\mu A$ for current reference; $1\,\mu A$ through pin V_{r1}). Note that to minimize the effect of ESD pad leakage and opamp bias current (about 1 nA worst-case), the current in R_1 is relatively large ($10\,\mu A$). If the reference voltage V_R were integrated on chip, this current would be negligible. The current consumption can also be strongly reduced by using a less accurate, or perhaps a duty-cycled, comparator. Since the aim of this chip was to investigate the feasibility of the proposed concept, it was optimized for accuracy rather than for low current consumption. In a fully integrated version, the reduced current through R_1 will compensate, at least partially, for the extra current required by the voltage reference and the temperature compensation circuitry [18,19].

---

[5]The area labelled as "capacitors" in Fig. 4.10 also contains also transistors M_1 and M_3 of the current reference, which are required to match MOS capacitors C_A and C_B.

Fig. 4.10 Die micrograph of the test chip

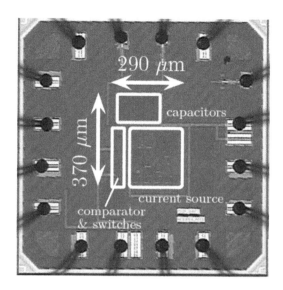

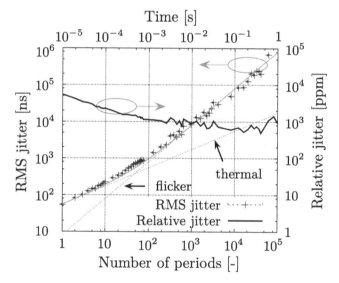

Fig. 4.11 Measured long-term jitter vs. time; extrapolated thermal and flicker noise components and their cumulative contribution are also plotted

Long-term jitter measurements for an output frequency of 100 kHz are reported in Fig. 4.11, together with lines showing the extrapolated thermal and flicker noise components. Period jitter is 52 ns (rms) and is dominated by comparator thermal noise. After a large number of periods, jitter is dominated by flicker noise from the current reference. As proven in Sect. 3.2.2, relative jitter [defined in (3.10)] becomes flat for increasing time for oscillators affected by thermal and flicker noise,

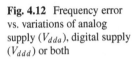

Fig. 4.12 Frequency error vs. variations of analog supply (V_{dda}), digital supply (V_{ddd}) or both

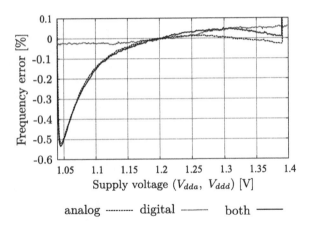

as observed in the measurements [20]. The relative jitter is 0.1% after one second and is negligible compared to the temperature-induced frequency drift.

Frequency pushing is shown in Fig. 4.12. The nominal supply voltage of the circuit should be 2.5 V (with pMOS and nMOS threshold voltages of 0.63 V and 0.57 V respectively) but the chosen topologies of the current reference and comparator allow functionality down to 1.05 V. The upper bound of the supply voltage is limited to 1.39 V by the start-up circuit in the current reference. With reference to Fig. 4.12, V_{dda} supplies the current reference, while V_{ddd} supplies the logic and the comparator. The increase of frequency with V_{ddd} is due to a decrease in the delay of the comparator, which is related to the period of $RESET$ in Fig. 4.9 and is fixed by logic circuitry. The supply voltage of the comparator can be increased up to 1.5 V (not shown in Fig. 4.12) without affecting its functionality and it can be adjusted to shift the input common-mode range of the comparator. Figure 4.12 shows that at an output frequency of 100 kHz, the error is less than 0.1% for supply voltages above of 1.12 V. The resulting supply sensitivity of 0.4%/V is comparable to most of the references reported in Table 3.1, but it is still higher than the state-of-the-art in the same table. This is due to the operation with a supply voltage (1.2 V) which is much lower than the nominal supply voltage (2.5 V).

Measurements on 11 samples from one batch were performed over the industrial temperature range (−40 to +85°C) using a temperature-controlled oven. The measurement setup was designed to accurately stabilize the temperature of the samples during measurements (to within 0.01°C); however, this stable temperature could only be set with an inaccuracy in the order of 0.1°C. As the samples were not tested simultaneously, the measured data was post-processed to eliminate errors due to temperature mismatch. For each sample, frequency has been measured as a function of temperature. The data points were then interpolated to extract the values of frequency corresponding to a fixed set of temperatures. In Fig. 4.13, measurements are compared to simulations of the circuit of Fig. 4.6, where the solid line was obtained by using ideal models for all components except transistors M_1 and M_3. The measured output frequency shows the same temperature dependence

Fig. 4.13 Output frequency measurements and average on 11 samples ($V_R = 0.2$ V); the output frequency expected from simulation of the current reference is also shown with the *solid line*

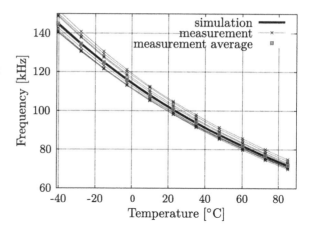

Fig. 4.14 Log–log plot of the output frequency (11 samples and their average for $V_R = 0.2$ V) vs. the absolute temperature; the best fit using the function $f(T) = f_0 T_\mu^\alpha$ is included for comparison

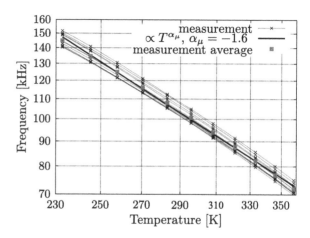

of mobility as the simulations and an untrimmed inaccuracy of 7% (3σ) at room temperature. Its temperature dependence is approximately proportional to T_μ^α, where T is the absolute temperature. The output frequency of the measured samples and the best fit of its average with the function T_μ^α (obtained for $\alpha_\mu = -1.6$) are shown in the logarithmic plot of Fig. 4.14. After one-point calibration, the frequency spread with respect to average frequency has been computed in Matlab from the data in Fig. 4.13 and it is below 1.1% (3σ) over the range from -22 to 85°C (Fig. 4.15). During the oven measurements, capacitors C_A and C_B were biased in deep inversion to reduce the effect of their spread, as explained in Sect. 4.4.4 ($V_{ddd} = 1.5$ V, $V_{r1} = 1.6$ V, $V_{r2} = 1.2$ V). This minimized the effect of threshold voltage spread on their capacitance, and ensured that the measured spread was mainly due to the core

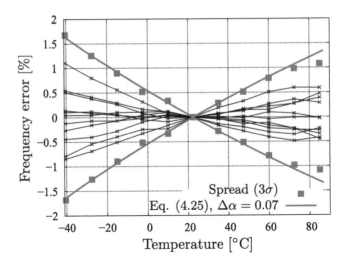

Fig. 4.15 Frequency error with respect to average frequency vs. temperature after one-point trimming at room temperature with $V_R = 0.25$ V for 11 samples

circuit. The model expressed by (4.25) is also shown in Fig. 4.15, demonstrating good agreement between the model and the experimental data. More details about the analysis of the residual spread are presented in Appendix B.

4.7   On Temperature Compensation

In the previous section, it has been shown that the output frequency of the mobility-based oscillator is strongly temperature dependent (Fig. 4.13). However, Fig. 4.15 shows that the spread in this temperature dependence is in the order of 1%. This will be the residual error in the output frequency if the temperature compensation scheme is perfect and if the oscillator is trimmed at a single temperature.

The compensation can be performed by varying a physical parameter in the oscillator circuit (Fig. 4.6), such as the voltage V_R, the gain of the current mirrors $M_1 - M_A$ and $M_1 - M_B$, or the capacitance of C_A and C_B. Alternatively, compensation can be introduced in the processing of the output frequency, for example by varying the multiplication factor of a cascaded frequency multiplier or, if an alarm signal after a fixed time period is needed (see Chap. 2), changing the number of reference periods to be counted in the fixed period. In all these schemes, the compensation parameter should be varied as a function of the temperature and so a temperature measurement error will lead to additional spread.

Fig. 4.16 Spread of the error of the compensated frequency vs. the standard deviation of the error in temperature estimation; simulation results and computation of (4.31) are plotted respectively with *lines* and *squares*

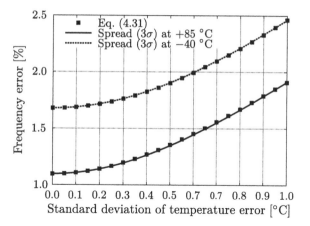

If the frequency is approximated as $f = f_0 T^{\alpha_\mu}$ in a limited temperature range, an error in the temperature measurement will cause a relative error in the compensated frequency given by

$$\sigma_{\frac{\Delta f}{f}} \approx \sqrt{\sigma_0^2 + \left|\frac{T}{f}\frac{\partial f}{\partial T}\right|^2 \sigma_{\frac{\Delta T}{T}}^2} = \sqrt{\sigma_0^2 + \alpha_\mu^2 \sigma_{\frac{\Delta T}{T}}^2} \qquad (4.31)$$

where $\sigma_{\frac{\Delta f}{f}}$ is the standard deviation of the error in the compensated frequency, σ_0 is the standard deviation of the error in the uncompensated frequency (i.e. the one reported in Fig. 4.15) and $\sigma_{\frac{\Delta T}{T}}$ is the standard deviation of the relative error in the temperature measurement. Simulations have been performed, assuming an ideal compensation by a multiplicative factor and a random error in temperature measurement, and the results are shown in Fig. 4.16. Equation (4.31) is also plotted in the figure using[6] $\alpha_\mu = -1.4$ for $-40°C$ and $\alpha_\mu = -1.9$ for $85°C$. An increase in the spread of less than 0.05% is observed at both extremes of the temperature range for a temperature sensing error with standard deviation of 0.2°C.

A lower sensitivity to temperature errors can be obtained if α_μ in (4.31) is decreased. This can be achieved by making V_R a temperature-*dependent* voltage instead of a temperature-independent voltage. A particularly interesting case is when V_R is proportional to the absolute temperature (PTAT). This would be easy to realize, since such voltages are commonly employed in bandgap voltage references [21]. The use of a PTAT V_R would result in $\alpha_\mu \approx 0.5$ [with reference to (4.27)] and, consequently, to a smaller spread for a fixed accuracy of the temperature sensor. A PTAT V_R has been applied to the test chip and results are reported in Fig. 4.17. To reduce measurement time, the behaviour of the circuit at arbitrary

---

[6]The values used for α_μ were obtained at $-40°C$ and $85°C$ from the slope of the average frequency characteristic in Fig. 4.14.

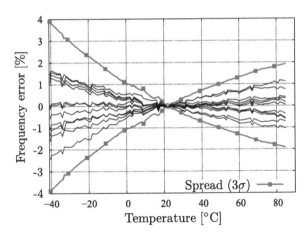

Fig. 4.17 Frequency error with respect to average frequency vs. temperature after one-point trimming at room temperature with V_R proportional to absolute temperature (PTAT)

values of temperature and voltage V_R were obtained by interpolating between actual measured data and the effect of compensation has been computed. The use of interpolation is the cause of the disturbances visible in Fig. 4.17. The application of a PTAT V_R increases the spread with respect to the constant V_R case (i.e. Fig. 4.15). Note that the compensation has been performed without adding any error in the temperature measurement. The larger spread can be explained by analyzing the effect of threshold voltage mismatch between transistors M_1 and M_3 in Fig. 4.6. Taking into account a threshold voltage mismatch ΔV_{th} between M_1 and M_3, (4.26) is modified as

$$f(T) = \mu_n k \frac{C_{ox} \frac{W_1}{L_1}}{C} \frac{(V_R(T) + \Delta V_{th})^2}{V_{r1} - V_{r2}} \tag{4.32}$$

After trimming at temperature T_0, it can be shown that the frequency error due to the threshold voltage mismatch is zero in the case of temperature-independent V_R and given by the following expression in the case of a PTAT V_R:

$$\frac{\Delta f'}{f'} \approx \frac{2\Delta V_{th}}{V_R(T_0)} \frac{T_0 - T}{T} \tag{4.33}$$

where f' is the output frequency after trimming. It can then be concluded that better performance can be achieved with compensation schemes which keep V_R temperature independent, and that the use of temperature-dependent V_R is only indicated if the matching between M_1 and M_3 can be significantly improved.

4.8   Effects of Packaging and Process Options

To explore the robustness of the mobility-based references, the oscillator has also been by implemented with both thin-oxide and thick-oxide MOS transistors in a 0.16-μm CMOS process, and samples packaged in both ceramic and plastic packages have been tested [22]. Compared to implementation in deep-submicron technology, the use of more mature technology can drastically reduce costs. Moreover, the gate leakage of thin-oxide transistors in deep-submicron technology reduces the accuracy of ultra-low-power circuitry, whereas it is negligible for both the thin and thick-oxide transistors in the chosen 0.16-μm process, thus allowing both options to be explored in the same process.

The implemented oscillator has the same architecture shown in Fig. 4.6. To improve the accuracy, the design of the current reference was slightly modified with respect to that in Fig. 4.8. The current source I_0 in Fig. 4.6 is realized on-chip by the bias circuit in the dashed box in Fig. 4.18. This consists of thin-oxide transistors and is supplied by $V_{dd2} = 1.8$ V. With the switches configured as shown in the figure, the opamp forces the voltage V_R across $R_1 = R_2 = 750$ kΩ to generate $I_0 = V_R/R_1$. This current is copied by current mirrors $M_7 - M_8$ and $M_5 - M_6$ and by flowing through R_2 generates a voltage difference between the gates of M_1 and M_3 equal to $R_2 I_0 = \frac{R_2}{R_1} V_R$. The mismatch of the resistors and the current mirrors together with the offset of the opamp introduces errors in V_R and consequently an error Δf in the output frequency. By periodically toggling the position of the switches and chopping the opamp, this mismatch-induced frequency error can be averaged out.

The mobility reference was implemented in a baseline SSMC 0.16-μm CMOS process. In order to test the robustness of the mobility-based reference to process options, two versions were made, one in which the current reference (Fig. 4.18) and MOS capacitors C_A and C_B in Fig. 4.6 were implemented with thin-oxide and one in which thick-oxide transistors were used. Since the comparator's delay may limit the reference's accuracy at high output frequencies, a third "slow" thin-oxide reference was implemented with $3\times$ larger oscillator capacitors. The thin-oxide, thick-oxide and slow thin-oxide references each occupy 0.06 mm^2 (Fig. 4.19) and their current consumption at room temperature is, respectively, 11.8 μA, 12.4 μA and 10.9 μA from a 1.2-V supply (V_{dd}) and 2.1 μA from a 1.8-V supply (V_{dd2}). All reference voltages (V_{r1}, V_{r2}, V_R) were generated externally. The samples have been packaged both in stress-free ceramic packages and in standard plastic packages without any stress-relieving coating.

The average output frequency of the various time references is shown in Fig. 4.20. At room temperature, the reference current is $I_1 = 200$ nA ($V_R = 0.275$ V) for the thin-oxide reference and $I_1 = 200$ nA ($V_R = 0.225$ V) for the thick oxide reference. For both references, $C_A = C_B \approx 7$ pF, $V_{r1} = 1.2$ V and $V_{r2} = 0.8$ V. These reference voltages were used for all the reported measurements. The thin-oxide and thick-oxide references only operate correctly over a limited temperature range, because the comparator's delay becomes significant for frequencies above 200 kHz and because the transistor's threshold voltage approaches the supply voltage at low temperatures. The measurements shows that the parameter α_μ in (4.23) is approximately equal to -1.7 and -1.9 for the thin-oxide and the thick-oxide references, respectively.

Measurements were made on 12 samples in ceramic packages and 12 samples in plastic packages from one batch. Over the military temperature range, the lowest

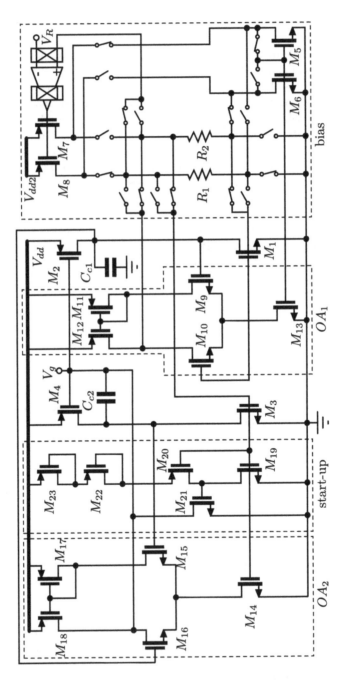

Fig. 4.18 Schematic of the current reference employed in the 0.16-μm CMOS test chip; the cascode transistors in series with $M_5 - M_8$ are omitted for clarity

Fig. 4.19 Die micrograph of the 0.16-μm CMOS test chip

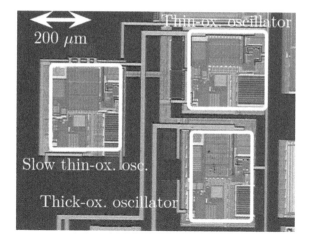

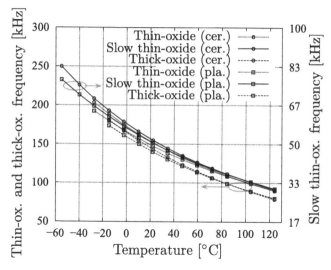

Fig. 4.20 Output frequency of the 0.16-μm time references

frequency spread of 1% (3σ) was achieved by the "slow" thin-oxide reference in ceramic packaging (Fig. 4.21). To easily compare the different results, the spread model of (4.25) has been least-square fitted to the observed 3σ-spread. Good agreement is observed between the model of (4.25) and the experimental data. The resulting 3σ values for $\Delta\alpha$ are reported in Fig. 4.21. While similar results were achieved by both thin-oxide references (as expected since they contain the same current reference), the spread of the thick-oxide reference is approximately 50% larger. Even with the limited number of available samples, the use of plastic packaging clearly results in more spread (about 2× more) than the use of ceramic packaging.

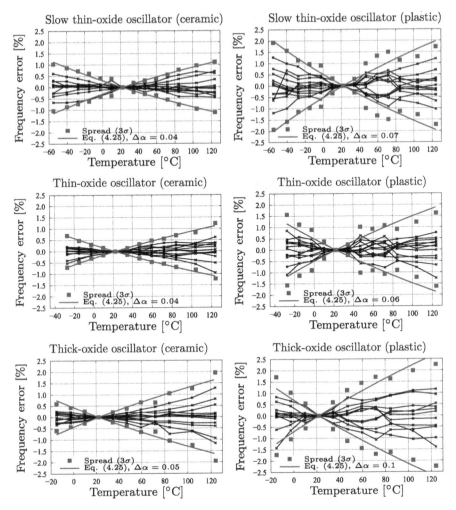

Fig. 4.21 Frequency error with respect to the average frequency vs. temperature after one-point trimming at room temperature for 12 samples of the 0.16-μm CMOS test chip (11 samples tested for the ceramic-packaged slow thin-oxide oscillator)

4.9   Conclusions

A fully integrated 65-nm CMOS mobility-based 100-kHz time reference has been presented. Its frequency inaccuracy, due to temperature, supply variations and noise, respectively, is 1.1% (3σ) from −22 to 85°C, 0.1% with a supply variation of 0.27 V and 0.1% (rms) over a one second time span. This shows that, by adopting an appropriate temperature compensation scheme, the electron mobility can be used to generate a reference frequency accurate enough for WSN applications and that the proposed architecture is both low-voltage and low-power, as required by autonomous sensor nodes.

Mobility-based time references can be also implemented in different processes and with different packaging. In a given process, their accuracy will depend both on the devices used and on the selected packaging. By using thin-oxide transistors and ceramic packaging in a 0.16-μm CMOS process, inaccuracies as low as 1% over the military temperature range can be achieved. Even when accuracy must be sacrificed for the sake of cost, and thus low-cost plastic packages are used, the resulting inaccuracy can be kept below 2% over the same temperature range. This demonstrates the robustness of the proposed references and their potential for low-cost application in low-power low-voltage integrated systems.

The required temperature compensation must have an accuracy of $0.2°C$ (1σ). While such level of accuracy has been reached in mature technologies with conventional bandgap temperature sensors at a power dissipation in the order of some tens of μW [18, 19], temperature sensors implemented in deep-submicron technologies, such as 65 nm [23] and 32 nm [24], achieve an accuracy of only a few degrees if special care is not taken. Since temperature compensation is a crucial point to demonstrate the viability of the mobility-based approach for WSN references, deep-submicron temperature sensors and temperature compensation strategies will be investigated in the following chapter.

References

1. Blauschild R (1994) An integrated time reference. ISSCC Dig. of Tech. Papers, pp 56–57
2. Tsividis Y (2003) Operation and modeling of the Mos transistor, 2nd edn. Oxford University Press, New York, p 12
3. Tsividis YP (1994) Integrated continuos-time filter design – an overview. IEEE J Solid State Circ 29(3):166–176
4. Sansen W, Op't Eynde F, Steyaert M (1988) A CMOS temperature-compensated current reference. IEEE J Solid State Circ 23(3):821–824
5. Jeon D, Burk D (1989) MOSFET electron inversion layer mobilities-a physically based semi-empirical model for a wide temperature range. IEEE Trans Electron Dev 36(8):1456–1463. DOI 10.1109/16.30959
6. Ghani T, Mistry K, Packan P, Thompson S, Stettler M, Tyagi S, Bohr M (2000) Scaling challenges and device design requirements for high performance sub-50 nm gate length planar CMOS transistors. In: 2000 Symposium on VLSI Circuits Dig. Tech. Papers, pp 174–175. DOI 10.1109/VLSIT.2000.852814
7. Lo SH, Buchanan D, Taur Y, Wang W (1997) Quantum-mechanical modeling of electron tunneling current from the inversion layer of ultra-thin-oxide nMOSFET's. IEEE Electron Dev Lett 18(5):209–211. DOI 10.1109/55.568766
8. Mukhopadhyay S, Neau C, Cakici RT, Agarwal A, Kim CH, Roy K (2003) Gate leakage reduction for scaled devices using transistor stacking. IEEE Trans VLSI Syst 11(4):716–730
9. O'Halloran M, Sarpeshkar R (2004) A 10-nW 12-bit accurate analog storage cell with 10-aA leakage. IEEE J Solid State Circ 39(11):1985–1996. DOI 10.1109/JSSC.2004.835817
10. Wang TJ, Ko CH, Chang SJ, Wu SL, Kuan TM, Lee WC (2008) The effects of mechanical uniaxial stress on junction leakage in nanoscale CMOSFETs. IEEE Trans Electron Dev 55(2):572–577. DOI 10.1109/TED.2007.912363
11. Sedra AS, Smith KC (1998) Microelectronics circuits, 4th edn. Oxford University Press, New York

12. Tsividis Y (2003) Operation and modeling of the MOS transistor, 2nd edn. Oxford University Press, New York, NY
13. Sebastiano F, Breems L, Makinwa K, Drago S, Leenaerts D, Nauta B (2009) A low-voltage mobility-based frequency reference for crystal-less ULP radios. IEEE J Solid State Circ 44(7):2002–2009
14. Annema AJ, Nauta B, van Langevelde R, Tuinhout H (2005) Analog circuits in ultra-deep-submicron CMOS. IEEE J Solid State Circ 40(1):132–143. DOI 10.1109/JSSC.2004.837247
15. Gregor R (1992) On the relationship between topography and transistor matching in an analog CMOS technology. IEEE Trans Electron Dev 39(2):275–282
16. Tuinhout H, Vertregt M (1997) Test structures for investigation of metal coverage effects on mosfet matching. In: Proceedings of IEEE International Conference on Microelectronic Test Structures, ICMTS 1997, pp 179–183. DOI 10.1109/ICMTS.1997.589386
17. Tuinhout H, Vertregt M (2001) Characterization of systematic MOSFET current factor mismatch caused by metal CMP dummy structures. IEEE Trans Semicond Manuf 14(4):302–310. DOI 10.1109/66.964317
18. Bakker A, Huijsing J (1996) Micropower CMOS temperature sensor with digital output. IEEE J Solid State Circ 31(7):933–937
19. Aita AL, Pertijs MA, Makinwa KAA, Huijsing JH (2009) A CMOS smart temperature sensor with a batch-calibrated inaccuracy of $\pm 0.25°C$ (3σ) from -70 to 130 °C. In: ISSCC Dig. of Tech. Papers
20. Liu C, McNeill J (2004) Jitter in oscillators with 1/f noise sources. Proc ISCAS 1:I–773–6. DOI 10.1109/ISCAS.2004.1328309
21. Meijer G, Wang G, Fruett F (2001) Temperature sensors and voltage references implemented in CMOS technology. IEEE Sensor J 1(3):225–234. DOI 10.1109/JSEN.2001.954835
22. Sebastiano F, Breems L, Makinwa K, Drago S, Leenaerts D, Nauta B (2011) Effects of packaging and process spread on a mobility-based frequency reference in 0.16-μm CMOS. In: Proceedings of ESSCIRC, pp 511–514
23. Duarte D, Geannopoulos G, Mughal U, Wong K, Taylor G (2007) Temperature sensor design in a high volume manufacturing 65nm CMOS digital process. In: Proceedings of IEEE Custom Integrated Circuits Conference (CICC), pp 221–224. DOI 10.1109/CICC.2007.4405718
24. Lakdawala H, Li Y, Raychowdhury A, Taylor G, Soumyanath K (2009) A 1.05 V 1.6 mW 0.45 °C 3σ-resolution $\Sigma\Delta$-based temperature sensor with parasitic-resistance compensation in 32 nm CMOS. IEEE J Solid State Circ (12):3621–3630

Chapter 5
Temperature Compensation

5.1 Introduction

In Chap. 4, it has been shown that after a single-point trim at room temperature the output frequency of mobility-based oscillators is characterized by a strong temperature dependence, which is larger then ±30% over the commercial temperature range from −40 to +85°C. However, if an ideal temperature compensation is applied, their inaccuracy is in the order of 1% over the same temperature range. As has been shown in Chap. 2, such inaccuracy is low enough for a large variety of applications, including WSN nodes. This chapter discusses how to keep such level of inaccuracy when going from an ideal to a practical temperature compensation scheme.

As shown in Chap. 4, due to the large temperature dependence of electron mobility, an error in temperature sensing of 1°C causes an additional frequency spread of the compensated reference of almost 1%. A large part of this chapter is therefore devoted to the implementation of a highly accurate temperature sensor in deep-submicron CMOS. After a brief description of the time reference architecture (Sect. 5.2), the requirements for the temperature sensor are sketched in Sect. 5.3 and the proposed sensor is introduced in Sect. 5.4. After a review of the sources of inaccuracies of a deep-submicron CMOS temperature sensor and of the solutions to circumvent them (Sect. 5.5), the selected circuit implementation is described in details in Sect. 5.6. The complete proposed temperature compensation scheme is then outlined in Sect. 5.8 and its experimental validation is described in Sect. 5.9.

5.2 Architecture of the Time Reference

The core of the proposed time reference is the current-controlled relaxation oscillator described in Chap. 4, in which the current is proportional to the electron mobility. As a result, its output frequency has the same temperature dependency as the electron mobility and so temperature compensation is needed. This can be

F. Sebastiano et al., *Mobility-based Time References for Wireless Sensor Networks*,
Analog Circuits and Signal Processing, DOI 10.1007/978-1-4614-3483-2_5,
© Springer Science+Business Media New York 2013

implemented either in an analog or in a digital manner. Analog compensation can be realized by adjusting one of the many analog control "knobs" available in the oscillator. Several tuning mechanisms have been listed in Sect. 4.7. For example, the control current of the oscillator can be implemented as the sum of the mobility-dependent current and a current with a complementary temperature dependency. This would be fairly easy if a physical effect with such a complementary temperature dependence had been identified. However, it has not and the generation of a suitable current would increase circuit complexity and be subject to errors which would lower the accuracy of the time reference. Moreover, Sect. 4.7 shows that the oscillator's spread increases if a non-negligible temperature dependence is added to one of its reference voltages.

Digital temperature compensation schemes do not suffer from these issues. They can be easily implemented by dividing or multiplying the oscillator frequency by a temperature-dependent factor. Operations in the time domain (frequency multiplication and division) are well suited to deep-submicron implementation. Errors in frequency multiplication and division derive mainly from delays in the circuit blocks, which can be negligible in a deep-submicron process thanks to the high speed of the integrated components. Moreover, in the case of a WSN node, a large part of the compensation can be implemented by re-using circuit blocks already present in the node. This minimizes circuit overhead and, consequently, power consumption.

Figure 5.1 shows the proposed compensation schemes. In Fig. 5.1a, the output frequency of the mobility-based oscillator f_{osc} is used as the reference of an integer-N Phase-Locked Loop (PLL), composed of a phase detector (PD), a loop filter, a voltage-controlled oscillator (VCO) and a divider. Via a pre-determined compensation curve, the digital output of a temperature sensor is mapped to the divider factor N_{mul}, in such a way that the output frequency f_{RF} remains constant over temperature. In Fig. 5.1b, f_{out} is obtained by direct division of f_{osc} by a temperature-dependent N_{div}. As already mentioned in Chap. 2, a WSN node needs both a low-frequency reference for time synchronization of the network protocol and a high-frequency reference for RF communication. Thus, most of the blocks for temperature compensation in Fig. 5.1, i.e. the PLL to generate f_{RF} and the divider to generate f_{out}, are already present in the WSN node. The addition of a non-linear mapping function would then be enough to realize both the system shown in Fig. 5.1 and to obtain both the high-frequency and low-frequency temperature-independent references.

In order to simplify hardware implementation, it would be convenient for both N_{mul} and N_{div} to be integers. To make the effect of temperature compensation negligible in the total accuracy, the quantization error due to the temperature compensation should be low enough, which can be expressed, for N_{mul} and N_{div} integers, as

$$N_{mul}, N_{div} > N_{res} \qquad (5.1)$$

$$N_{res} = \left\lfloor \frac{10}{a} \right\rfloor \qquad (5.2)$$

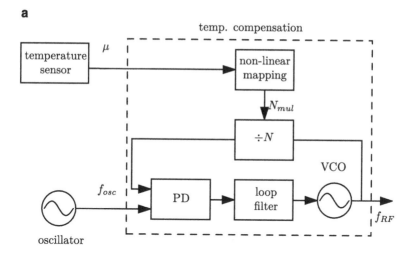

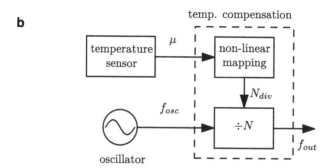

Fig. 5.1 Architectures for the time reference with temperature compensation implemented by means of frequency multiplication (**a**) and frequency division (**b**)

where N_{res} is the minimum factor needed to achieve the required resolution, a is the aimed accuracy for the whole reference and a factor 10 has been assumed between the reference accuracy and the quantization error of the compensation scheme, to make the latter negligible. Considering that $N_{mul} = \frac{f_{RF}}{f_{osc}}$ and $N_{div} = f_{osc} T_{synch}$ with $T_{synch} = f_{out}^{-1}$, the following inequality must be satisfied

$$\frac{f_{RF}}{\left\lfloor \frac{10}{a} \right\rfloor} > f_{osc} > \frac{\left\lfloor \frac{10}{a} \right\rfloor}{T_{synch}} \tag{5.3}$$

It has been proven in Chap. 2 that WSNs can tolerate $a \approx 1\%$, while adopting an RF frequency of the order of 1 GHz and using time spans between 10 and 100 ms for the network synchronization. It follows from (5.3) that the oscillator frequency must be in the range between 1 MHz and 10 kHz, while keeping at least 10 bits of resolution for the divider ratios. A value of $f_{osc} = 100$ kHz is then an appropriate choice. Since the accuracy of the proposed mobility-based reference can be demonstrated by any of the two systems of Fig. 5.1, only the system in Fig. 5.1b has been implemented and it is discussed in the following.

5.3 Requirements on the Temperature Sensor

5.3.1 Sampling Rate

The required sampling rate of the temperature sensor depends on the thermal behavior of the packaged time reference. A thermal model of the time reference using an approximated lumped-element equivalent circuit is shown in Fig. 5.2. The voltages at the circuit nodes represent the temperature at different physical points of the system: T_{die}, T_c and T_a are, respectively, the temperature of the die, the case, i.e. the external surface of the package, and the ambient. The currents represent the power (measured in W) that flows through the different parts. P is the power dissipated by the whole time reference on the die. The capacitors C_{die} and C_{pkg} (measured in J/K) represent the thermal capacitance of, respectively, the silicon die and the package. R_{jc} and R_{ca} (measured in K/W) are called respectively the *junction-to-case thermal resistance* and *case-to-ambient thermal resistance*.[1]

The values of the different components depend on the die and package specifications and on how the package is mounted (attached to a board, in still air, subjected to airflow, ...). However, practical estimates can be given. The die thermal capacitance can be computed as

$$C_{die} = c_{Si}\rho_{Si}At \tag{5.4}$$

where $c_{Si} = 700$ J/K/kg is the specific heat of silicon, $\rho_{Si} = 2{,}330$ kg/m^3 the density of silicon, A the die area and t the die thickness. For a 1-mm^2 die and a typical die thickness after grinding of 380 μm, $C_{die} = 620$ μJ/K. R_{jc} and R_{ca} vary typically in the range 1–100 K/W and 10–200 K/W, respectively, for packages without special power dissipation features and in still air [1, 2]. The thermal capacitance of typical packages is in the order of 0.1 J/K [3]. An estimate of its order of magnitude can also

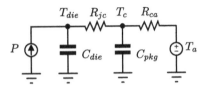

Fig. 5.2 Thermal model
of the time reference

---

[1]In package datasheets the *junction-to-ambient thermal resistance* is also often reported, which for the simple model presented is equal to $R_{ja} = R_{jc} + R_{ja}$.

be obtained noting that for the packaged samples described in Sect. 5.9, a thermal time constant of the order of 10 s was measured, resulting in $C_{pkg} \approx 0.1$ J/K for $R_{ca} \approx 100$ K/W. Note that for an average power dissipation $P \approx 100\,\mu$W, typical for WSN nodes as explained in Chap. 2, self-heating of the die is limited to a few tens of millikelvin.

The temperature of the die is mainly determined by the ambient temperature and the thermal dynamics of the system in Fig. 5.2. For the parameters listed above, ambient temperature variations are filtered with a cut-off frequency of the order of 0.1 Hz, partly due to the package. However, even if C_{pkg} were zero, the thermal cut-off would still be limited by C_{die} to about 1–10 Hz.

The conversion rate must be high enough to cope with the temperature variations expected in a particular application. A slow sensor might not be able to accurately track such variations and would thus introduce extra readout errors. Since self-heating can be neglected for a WSN node, temperature variations are only due to the environment. For typical WSN applications, such as environmental monitoring, environmental temperature variations are limited to a few degrees in a time span ranging from seconds to minutes. Moreover, any high-frequency components of temperature variations will be suppressed by the thermal network of Fig. 5.2. Consequently, the temperature behavior can be accurately tracked using a few temperature readings per second.

5.3.2 Accuracy

The sensor must be accurate enough to compensate for the mobility-based oscillator without worsening its intrinsic spread. A detailed analysis in Sect. 4.7 has shown that a standard deviation of the temperature error of less than 0.2°C (1σ) does not contribute significantly to this spread. It was shown that the inaccuracy introduced by a temperature reading error ΔT can be approximated as

$$\frac{\Delta f_{out}}{f_{out}} \approx \frac{\Delta T}{f_{osc}} \frac{\partial f_{osc}}{\partial T} = \alpha \frac{\Delta T}{T} \tag{5.5}$$

where the oscillation frequency is approximated as $f_{osc} = f_0 T^{\alpha}$. In the worst case, $\alpha = -1.9$ at 85°C (see Sect. 4.7) and the inaccuracy of the temperature sensor must be kept below approximately 0.2°C (1σ) to keep the frequency error due to the compensation limited to 0.1%.

5.3.3 Voltage Supply and Power Consumption

The supply specification for the oscillator (Chap. 4) can be applied to the compensated reference: the supply voltage should be as low as 1.2 V while achieving a supply sensitivity of less than 1%/V. Using the symbol $S_x^y = \frac{\Delta y/y}{\Delta x/x}$ for the *sensitivity factor* of parameter y with respect to the variation of variable x, the supply sensitivity ($\frac{\Delta f_{out}}{f_{out}}/\Delta V_{dd}$) can be expressed as a function of the temperature-sensor supply sensitivity factor $S_{V_{dd}}^{T}$ and the oscillator supply sensitivity factor $S_{V_{dd}}^{f_{osc}}$:

$$\frac{\frac{\Delta f_{out}}{f_{out}}}{\Delta V_{dd}} = \frac{1}{V_{dd}} \cdot S_{V_{dd}}^{f_{out}} = \frac{1}{V_{dd}} \cdot \left(S_{V_{dd}}^{f_{osc}} + \frac{T}{f_{osc}} \frac{\partial f_{osc}}{\partial T} S_{V_{dd}}^{T} \right) \tag{5.6}$$

With the approximations used in the previous section, it can be derived that the absolute value of the temperature-sensor supply sensitivity, i.e. $\Delta T/\Delta V_{dd}$ must be constrained by the following inequality:

$$\frac{\Delta T}{\Delta V_{dd}} = \frac{T}{V_{dd}} \cdot |S_{V_{dd}}^{T}| < \frac{f_{osc}}{V_{dd}} \cdot \left| \frac{\partial f_{osc}}{\partial T} \right|^{-1} \left(|S_{V_{dd}}^{f_{out}}| - |S_{V_{dd}}^{f_{osc}}| \right) = \tag{5.7}$$

$$= \frac{1}{V_{dd}} \cdot \frac{T}{|\alpha|} \left(|S_{V_{dd}}^{f_{out}}| - |S_{V_{dd}}^{f_{osc}}| \right) < 0.8 \, °C/V \tag{5.8}$$

where $|\alpha| = 1.9$ at 85°C and $S_{V_{dd}}^{f_{osc}}/V_{dd} = 0.6\%/V$ has been extracted from the measurement results of Fig. 4.12.

The power consumption of the total reference is limited to about 50 μW according to the specifications drawn in Chap. 2. Accounting for the power consumption of the oscillator of 41 μW (see Chap. 4), a budget of approximately 10 μW is available for the temperature sensor. This is possible under the assumption, verified later in this chapter, that the contribution of the frequency divider and the non-linear mapping block of Fig. 5.1b to the power budget is negligible.

5.4   Temperature-Sensor Operating Principle

CMOS temperature sensors with an inaccuracy of less than ±0.1°C over the military temperature range have been demonstrated in a mature technology (0.7-μm CMOS) [4, 5]. They are usually based on the bandgap principle, i.e. exploiting the temperature dependency of PNP transistors, and achieve high accuracy by employing a single-temperature trim as well as precision circuit techniques, such as offset cancellation, Dynamic Element Matching (DEM) and curvature correction.[2]

The same sensing principle has been employed in temperature sensors in 65 nm [6] and 32 nm [7], but these only achieve inaccuracies of about 5°C. This lack

---

[2]Throughout this chapter, the terms *batch-calibration* or *correction* (curvature correction, non-linear correction) refer to the procedure of estimating the average error of a production batch from the measurements of a limited number of samples from that batch, and by adjusting all individual samples in the same manner and by the same amount based on that estimate.

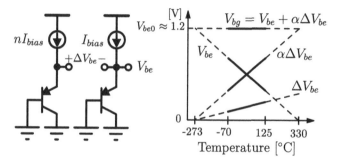

Fig. 5.3 Principle of operation of the temperature sensor and temperature dependence of voltages in the sensor core

of accuracy is partly due to the non-idealities of parasitic PNP transistors in deep-submicron technologies. Other sensing principles have been proposed for deep-submicron applications, such as the use of thermistors [8], the measurement of ring oscillator frequency [8] or of MOS-transistor leakage [9]. These approaches require either multi-temperature trimming, or suffer from inaccuracies of a few degrees Centigrade even over temperature ranges much narrower than the standard military or industrial temperature ranges. Sensors based on inverter delay have been proposed as good candidates for VLSI integration because of their compact layout. However, in a 0.35-μm CMOS prototype, a two-temperature trimming was necessary to achieve an inaccuracy of -0.4 to $+0.6°$C (3σ) over the range from 0 to 90°C [10]. Furthermore, the sensor's power supply sensitivity was quite high: about 10°C/V at room temperature, which is some two orders of magnitude worse than that of PNP-based sensors.

The sensing principle of a bandgap (or bipolar-transistor-based) temperature sensors is depicted in Fig. 5.3. The sensor's core consists of a pair of matched bipolar transistors (diode-connected PNPs) biased by two currents with ratio n, to produce two temperature-dependent voltages V_{be} and ΔV_{be}. The base-emitter voltage V_{be} of one transistor and the difference in base-emitter voltages ΔV_{be} can be approximated as

$$V_{be} \approx \frac{kT}{q} \ln \left(\frac{I_{bias}}{I_S} \right) \tag{5.9}$$

$$\Delta V_{be} = \frac{kT}{q} \ln \left(\frac{n I_{bias}}{I_S} \right) - \frac{kT}{q} \ln \left(\frac{I_{bias}}{I_S} \right) = \frac{kT}{q} \ln n \tag{5.10}$$

where k is the Boltzmann's constant, T is the absolute temperature, q is the electron charge, I_{bias} is the bias current and I_S is the saturation current of the transistor. ΔV_{be} is proportional to absolute temperature (PTAT) and independent of process and bias conditions.

ΔV_{be} can be fed to an analog-to-digital converter (ADC) to produce a digital temperature reading. As shown in Fig. 5.3, a PTAT digital output μ can then be generated by combining V_{be} and ΔV_{be} as follows [11]:

$$\mu = \frac{\alpha \Delta V_{be}}{V_{bg}} = \frac{\alpha \Delta V_{be}}{V_{be} + \alpha \Delta V_{be}} \tag{5.11}$$

and using the appropriate scale factor α ($\alpha = 18$ in this work). An output in degree Celsius can then be obtained by scaling:

$$D_{out} = A \cdot \mu + B \tag{5.12}$$

where $A \approx 600$ and $B \approx -273$ [11].

5.5 Sources of Inaccuracy

5.5.1 Non-idealities of Bipolar Transistors

CMOS temperature sensors are usually based on substrate PNPs [4–7,12]. As shown in Fig. 5.4a, these consist of a p+ drain diffusion (emitter), an n-well (base) and the silicon substrate (collector) and are available in most CMOS processes. Since the silicon substrate is usually tied to ground, the PNP must be biased via its emitter (Fig. 5.5a). While (5.9) is valid for this configuration under the approximation $I_C \approx I_{bias}$, it is possible to derive [11]

$$V_{be} = \frac{kT}{q} \ln\left(\frac{I_C}{I_S}\right) = \frac{kT}{q} \ln\left(\frac{I_E - I_B}{I_S}\right) = \frac{kT}{q} \ln\left(\frac{I_{bias}}{I_S} \frac{\beta}{\beta + 1}\right) \tag{5.13}$$

where I_E and I_B are emitter and base currents and $\beta \triangleq \frac{I_C}{I_B}$ is the current gain of the transistor. The finite current gain and its spread affect both the curvature and the spread of V_{be}. The additional curvature can be compensated for by using standard methods for V_{be}-curvature compensation (see Sect. 5.5.4), but the additional spread directly impacts the sensor's accuracy. As can be understood from (5.13), this effect is negligible for high β but becomes increasingly significant as β decreases [4].

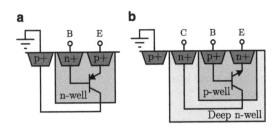

Fig. 5.4 Simplified cross section of (**a**) a substrate PNP and (**b**) a vertical NPN in CMOS technology

Fig. 5.5 Bipolar transistors configurations to generate V_{be} using (**a**) a substrate PNP and (**b**) a vertical NPN

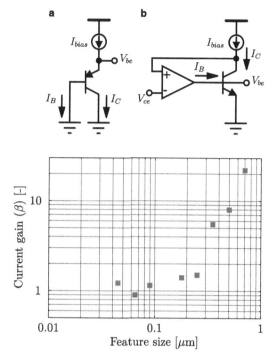

Fig. 5.6 Current gain β of substrate PNP transistors vs. the minimum gate length for various CMOS processes (data from [11] and several design manuals)

The current gain of the substrate PNPs available in several CMOS processes is reported in Fig. 5.6. It approaches unity in deep-submicron processes, making it difficult to implement accurate temperature sensors with these devices. As an alternative, parasitic NPN transistors can be employed, which can be directly biased via their collectors. Lateral NPN transistors in CMOS technology have been used in temperature sensors [13] but their $I_C - V_{be}$ characteristic deviates from (5.13) due to various extra non-idealities [11]. A better option is the vertical NPN [14, 15], which consists of an n+ drain diffusion (emitter), a p-well (base) and a deep n-well (collector), all standard features in deep-submicron processes (Fig. 5.4b). Their only disadvantage is a higher sensitivity to packaging stress compared to vertical PNPs: under stress condition typical of plastic package ($\approx \pm 150$ MPa), vertical NPNs shows a V_{be} variation of about 3 mV, which is equivalent to a temperature error of about 1°C; those variations are 60% smaller for vertical PNPs [16].

As shown in Fig. 5.5b, a vertical NPN can be accurately biased via its collector, while the required base current is provided by the feedback amplifier. The resulting base-emitter voltage will then be independent of the transistor's current gain. It should also be noted that this circuit can tolerate lower supply voltages than a diode-connected PNP. With reference to Fig. 5.5a, the PNP-based circuit requires a minimum supply voltage equal to the sum of V_{be} and the current source's headroom. Since V_{be} can be as high as ≈ 800 mV at the lower bound of the military temperature range (-55°C) and a certain headroom is required to ensure current source accuracy,

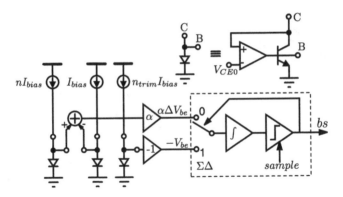

Fig. 5.7 Principle of operation of the charge-balancing converter

the minimum supply voltage can easily exceed 1.2 V. For the NPN-based circuit in Fig. 5.5b, however, the supply voltage primarily has to accommodate the sum of the NPN's saturation voltage $V_{ce} \approx 0.3\,\text{V} \ll V_{be}$ and the current source's headroom. Although it must also ensure the functionality of the branch, comprising the base-emitter junction and the amplifier, that supplies I_B, the accuracy of I_B is ensured by the feedback loop's gain and consequently the amplifier does not require much headroom. The minimum supply voltage can thus be lower than in the case of a diode-connected PNP. This is an advantage in deep-submicron designs, which must typically operate at supply voltages of 1.2 V or lower.

5.5.2 ADC Accuracy and Quantization Noise

The digital output μ in (5.11) can be obtained by connecting the bipolar core to the charge-balancing converter shown in Fig. 5.7 [17]. Here, a bias circuit generates a supply-independent current I_{bias}. Scaled copies of this current bias a pair of vertical NPNs at an $n{:}1$ collector current ratio and a third NPN with a current $n_{trim}I_{bias}$. The resulting voltages ΔV_{be} and V_{be} constitute the inputs of a first-order $\Sigma\Delta$ ADC. The ADC integrates $-V_{be}$ when the bitstream $bs = 1$ and integrates ΔV_{be} when $bs = 0$. Thanks to the negative feedback, the average input of the integrator is equal to zero, i.e. the integrated charge is balanced, which can be expressed as

$$(1 - \mu) \cdot \alpha \cdot \Delta V_{be} - \mu \cdot V_{be} = 0 \qquad (5.14)$$

where the bitstream average is $\mu = \langle bs \rangle$. From (5.14) it follows that the resulting μ satisfies (5.11).

In the practical implementation of the charge-balancing converter, a sensitive point is the implementation of the amplification factor α. An integer factor α is usually adopted, so that it can easily be realized by an array of α matched elements,

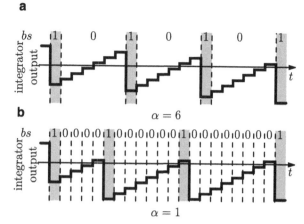

Fig. 5.8 Integrator output and output bitstream of a fragment of the temperature conversion of the system in Fig. 5.7 for different values of α; the *dashed lines* indicate the sampling of the comparator

e.g. capacitors [4]. The accuracy of α is therefore limited by the matching of these elements. That would require either a large chip area for the matched elements or the use of DEM techniques that would add to the complexity and area of the sensor.

Alternatively, the factor α can be realized by multiple integrations during the $bs = 0$ phase [18]. This is depicted in Fig. 5.8a for the case $\alpha = 6$. The α amplifier in Fig. 5.7 is removed from the system of Fig. 5.7 and when $bs = 0$, ΔV_{be} is integrated in $\alpha = 6$ successive cycles. When $bs = 1$, $-V_{be}$ is integrated in a single cycles. At the end of the $\alpha = 6$ cycles, the comparator's output is updated. With this solution, a single element can be used to implement $\alpha = 6$ and conversion speed is traded for accuracy. This is because α times more cycles are used to obtain an accurate multiplication factor. The comparator is sampled only after a single integration for the $bs = 1$ phase or after a series of α integrations for the $bs = 0$ phase. An additional improvement in resolution can be achieved if the comparator is sampled more rapidly, e.g. after every integration, as shown in Fig. 5.8b. Note that this is equivalent to multiple integrations with $\alpha = 1$. The effectiveness of adopting a smaller value for α has been demonstrated by Matlab simulation of the first-order $\Sigma\Delta$ converter for $\alpha = 18$ and $\alpha = 2$. The case $\alpha = 2$ is preferred over $\alpha = 1$ for ease of implementation, as will be shown in Sect. 5.6.4. The results are shown in Fig. 5.9, where the peak quantization error over the temperature range from -70 to $125°C$ is plotted vs. conversion time. In the simulation, the length of the different phases required by a practical circuit (see Sect. 5.6) has been used to obtain a practical comparison. Since $V_{be} \gg \Delta V_{be}$, the time needed for the integration of V_{be} is longer than that for ΔV_{be}. The adopted durations[3] of the $bs = 1$ phase and the $bs = 0$ phase are, respectively, $390\,\mu s$ and $100\,\mu s$ for $\alpha = 18$, and $70\,\mu s$ and $100\,\mu s$ for $\alpha = 2$.

---

[3]With reference to the symbols used in Sect. 5.6.4, for $\alpha = 18$, the $bs = 1$ phase is the same as in the case $\alpha = 2$, while the length in the $bs = 0$ phase has been assumed equal to $T_1 + (\alpha - 1)T_2 = 50\,\mu s + 17 \cdot 20\,\mu s = 390\,\mu s$.

Fig. 5.9 Simulated peak
quantization error over the
temperature range from -70
to $125°C$ vs. conversion time
for different values of α

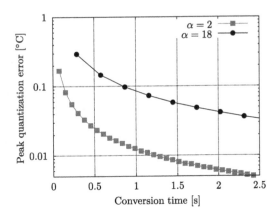

It should be noted that if $\alpha \neq 18$, then V_{bg} in (5.11) will no longer be temperature independent and the bitstream average μ will no longer be PTAT. If required, a digital back-end (similar to the one in [7]) can be used to compute a PTAT output, according to the relation

$$\mu_{PTAT} = \frac{\alpha_{PTAT} \cdot \mu}{\alpha + (\alpha_{PTAT} - \alpha)\mu} \tag{5.15}$$

where α_{PTAT} is the value required in (5.11) to obtain a PTAT output, and α is the value actually used in the charge-balancing converter.

It can be concluded that, for the same conversion time, using a smaller value of α results in lower quantization error, thanks to the increased granularity of the charge-balancing process. One drawback of this approach is the need for a digital back-end to implement the non-linear correction described by (5.15). However, in a deep-submicron CMOS technology, this requires little extra chip area and power dissipation. Additionally, another difference between this approach and that of adopting a larger α involves the amplitude of the signal handled by the $\Sigma\Delta$ converter. As shown in Fig. 5.10, the bitstream average μ, and, consequently, the signal converted by the ADC, is lower for a lower value of α. This is not an issue for a first-order $\Sigma\Delta$ ADC, since it is unconditionally stable and works properly for any input value in its input range, i.e. for $0 \leq \mu \leq 1$. However, higher-order $\Sigma\Delta$ converters suffer from overload for inputs at the edge of this range, i.e. for μ near 0 or 1. Thus, when using a low value of α, the choice of converter may be limited to first-order $\Sigma\Delta$ or MASH converters [19].

5.5.3   Process Spread

Since accurate current references are not available in CMOS, I_{bias} is derived by forcing a well-defined voltage, e.g. ΔV_{be}, across a resistor. However, due to the spread of this resistor and the spread of I_S, the V_{be} of the biased transistor will still

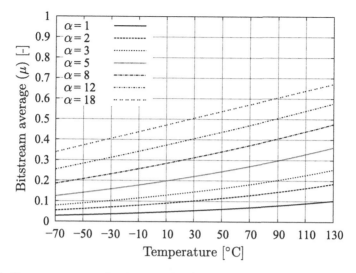

Fig. 5.10 Bitstream average value for α ranging from 1 to 18; in this example, for $\alpha = 18$, $\mu = \mu_{PTAT}$

spread. As shown in [20], this spread is PTAT in nature, and can be cancelled simply by trimming the bias current used to generate V_{be}, i.e. by trimming n_{trim} in Fig. 5.7 [4]. In this way, a single-point trim is enough to compensate for process spread.

The process spread affects also the accuracy of the bias-current ratio n of in Fig. 5.7 and, consequently, the accuracy of ΔV_{be} [see (5.10)]. This can be overcome by the application of the well-known DEM [11].

5.5.4 Non-linearity of V_{be}

In the previous sections, the temperature behavior of V_{be} has been considered to be linear. In practice, V_{be} shows a non-linearity mainly consisting of a second-order term [21]. Over the military temperature range, this can be as large as 1°C [4]. Since the output of the temperature sensor is fed to a non-linear mapping block (Fig. 5.1), any non-linearity could be taken into account when designing the non-linear mapping function. However, a temperature sensor with linear output can be more easily compared to the state-of-the-art. Moreover, having a linear temperature reading available on-chip can be convenient for different purposes in the WSN node. For those reasons, non-linearities corrections to provide a linear temperature reading are described in the following.

The non-linearity in μ can be compensated for by making the temperature coefficient of the denominator of (5.11), i.e. V_{bg}, positive [17]. This can be accomplished by slightly increasing I_{bias} compared to the value required to make V_{bg} temperature-independent. Any systematic residual non-linearity can then be compensated for by digital post-processing. A full conversion then consists of the following steps:

1. The charge-balancing converter is operated with $\alpha = 2$, as explained in Sect. 5.5.2.
2. The output bitstream bs is decimated to obtain

$$\mu = \frac{2\Delta V_{be}}{V_{be} + 2\Delta V_{be}} \qquad (5.16)$$

3. A PTAT ratio μ_{PTAT} is computed:

$$\mu_{PTAT} = \frac{9 \cdot \mu}{1 + 8 \cdot \mu} \qquad (5.17)$$

4. The residual non-linearity in μ_{PTAT} is compensated for with the help of a compensating polynomial.

5.6 Temperature-Sensor Circuit Description

A block diagram of the sensor is shown in Fig. 5.11 [22]. The circuit design of the bias circuit generating I_{bias}, the bipolar front-end and the $\Sigma\Delta$ are described in detail in the following sections, together with the choice of the bias currents for the bipolar core.

5.6.1 Current Level in the Bipolar Core

The bias currents of the NPN transistors in the bipolar core are constrained by several requirements, such as accuracy, noise and conversion speed. For low collector currents, the approximation $I_C \gg I_S$ used in (5.9) is not valid anymore and ΔV_{be} must be expressed as

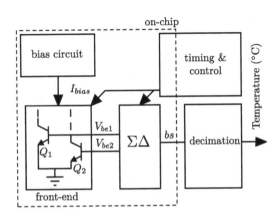

Fig. 5.11 Block diagram of the temperature sensor

Fig. 5.12 Maximum
temperature error over
the military range due
to spread in ΔV_{be} for
different bias current and bias
currents ratio n

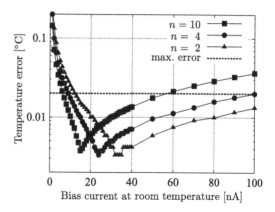

$$\Delta V_{be} = \frac{kT}{q} \ln \left(\frac{\frac{n I_{bias}}{I_S} + 1}{\frac{I_{bias}}{I_S} + 1} \right) = \frac{kT}{q} \ln n + \frac{kT}{q} \ln \left(\frac{1 + \frac{I_S}{n I_{bias}}}{1 + \frac{I_S}{I_{bias}}} \right) \qquad (5.18)$$

Thus, the bias current must be significantly larger than the saturation current in order to obtain an accurate PTAT voltage, especially at higher temperatures, since I_S increases rapidly with temperature and can reach pico-Ampere levels at 125°C.

For large bias currents, the accuracy of ΔV_{be} is impaired by the parasitic resistances R_B and R_E in series with the emitter and the base junction respectively. In this case ΔV_{be} may be expressed as

$$\Delta V_{be} = \frac{kT}{q} \ln n + R_B(I_{B1} - I_{B2}) + R_E(I_{E1} - I_{E2}) \qquad (5.19)$$

$$= \frac{kT}{q} \ln n + \left[\frac{R_B}{\beta} + R_E \left(\frac{1}{\beta} + 1 \right) \right] (n - 1) I_{bias} \qquad (5.20)$$

$$= \frac{kT}{q} \ln n + R_S(n - 1) I_{bias} \qquad (5.21)$$

where $I_{B1,2}$ and $I_{E1,2}$ are the base and emitter currents of $Q_{1,2}$ and R_S is the equivalent series resistance [11]. Typical values for R_B and R_E are in the order of 100 Ω and 10 Ω, respectively. Considering that the current gain β is commonly lower than ten in deep-submicron processes, R_S will be in the order of some tens of Ohms, leading to a non-negligible temperature error for bias currents higher than a few hundred nano-Amperes.

The additional terms in (5.18) and (5.21) make ΔV_{be} non-PTAT. Moreover, those terms will give rise to extra spread, due to the process spread of I_S, R_S and I_{bias}. Figure 5.12 shows the simulated effect of ΔV_{be} spread on the temperature reading for the NPN transistors available in the adopted technology with a $5 \times 5 \, \mu$m emitter area, with the added assumption that the spread in I_{bias} is ±20%. Since no accurate spread models for the parasitic resistances and saturation currents were available, these parameters were kept constant. The maximum allowable error (dashed line in Fig. 5.12) due to spread should be less than 10% of the target inaccuracy. It can be seen that several pairs of the design parameters n and I_{bias} meet those requirements. A larger n is preferable, because it implies a larger ΔV_{be} and consequently more relaxed requirements on the ADC. However, a larger bias current is also advantageous since it results in less noise and less sensitivity to other leakage currents [23]. Based on these considerations, $n = 4$ and $I_{bias} = 50\,\text{nA}$ at room temperature have been conservatively chosen.

5.6.2 Bias Circuit

In the bias circuit (Fig. 5.13), transistors Q_a and Q_b are biased by a low-voltage cascode mirror (A_1 and $M_1 - M_4$) with a 2:1 current ratio, forcing a PTAT voltage across polysilicon resistor $R_E = 180\,\text{k}\Omega$ and making the emitter current I_E of Q_b supply-independent. The gates of the cascode transistors $M_{3,4}$ are tied to ground ($V_{casc} = 0\,\text{V}$). The cascade of A_2 and M_{11} provides the base currents for Q_a and Q_b in a configuration similar to that shown in Fig. 5.5b. The bias current $I_{bias} = I_E$ can be derived by generating and summing copies of the collector current I_C and the base current I_{Bb} of Q_b. If the current gains β of Q_a and Q_b were equal and consequently $I_{ba} = 2I_{Bb}$ held for their base currents, I_{bias} could be obtained by mirroring the drain current of M_{11} with a gain of 1/3 and adding it to a copy of I_C. However, since β is a (weak) function of collector current, a replica circuit is used to bias the matched transistor Q_c with the same collector current of Q_b and obtain an accurate copy of I_{Bb} through M_{12}. Copies of I_{Bb} and I_C (through M_{13} and M_7) are then summed at the input of a low-voltage current mirror [24].

Unlike PNP-based bias circuits [4, 5, 12], the circuit in Fig. 5.13 does not need low-offset amplifiers. This is because the loop comprising the base-emitter junctions of $Q_{a,b}$ and resistor R_E can be directly realized with NPNs but not with substrate PNPs. In the presented circuit, the function of the feedback amplifiers and the low-voltage current mirror is only to equalize their collector-base voltages. Thus, their offset specifications are relaxed. However, since the base currents are relatively large (due to $\beta < 5$), and the transconductance of A_2 and A_3 is rather low, the common-source buffers M_{11} and M_{12} are used to reduce the voltages present at the input of A_2 and A_3. The collector voltage is set to V_{CE0}, obtained by biasing R_{CE} with a copy of I_C.

A_2 and A_3 are implemented as current-mirror-loaded pMOS differential pairs with tail currents of 340 nA at room temperature and current-mirror loads. Their respective feedback loops are stabilized by Miller capacitors $C_{c1,2}$ and the associated zero-cancelling resistors $R_{c1,2}$. A_1 is a current-mirror OTA [25] with a pMOS input pair. The associated feedback loop is stabilized by Miller capacitor C_{c3}. This is kept reasonably small (1 pF), by using a low bias current (8 nA) combined with a mirror attenuation of ten to keep the OTA's effective transconductance low. The bias currents of the amplifiers are scaled copies of I_C and are thus approximately PTAT and supply-independent.

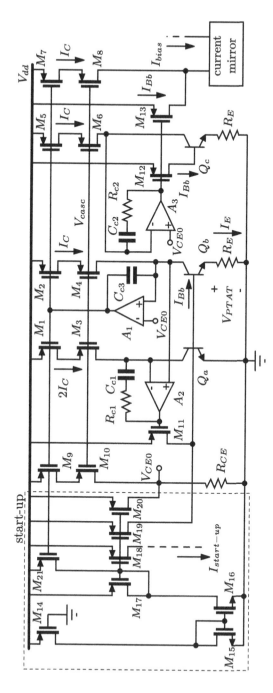

Fig. 5.13 Schematic of the bias circuit

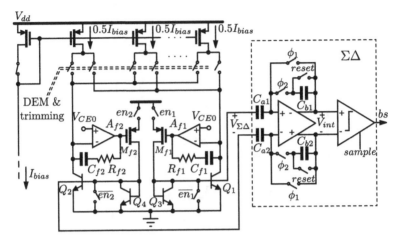

Fig. 5.14 Schematic of the bipolar core and of the $\Sigma\Delta$ ADC

Thanks to the use of NPNs and of the feedback loops, the circuit is able to work at low supply voltages and low temperatures. Simulations show that, for the adopted process, the effect of supply variations on I_{bias} is less than 300 ppm/V down to a supply voltage of 1.2 V at $-70°C$ (for which $V_{be} > 800$ mV).

Due to the self-biasing nature of the circuit, a start-up circuit is required. The long transistor M_{14} generates a current I_{D14} in the order of few tens of nA, which is lower than the I_C at the correct operation point for any operating condition and process corner. This current is compared to I_C by the current comparator comprising M_{16}, M_{17} and M_{21}. If I_{D14} is larger than I_C, i.e. if the circuit has not yet started up, the difference $I_{D14} - I_C$ is mirrored by $M_{17} - M_{20}$ and used to start-up the circuit. The start-up current is delivered to the bases of $Q_{a,b}$, to resistor R_{CE} and as bias currents of $A_{1,2,3}$ (not shown in the schematic), which would otherwise be off because of the low I_C. The injection of $I_{start-up}$ causes the currents in all the branches to increase and reach the stable operation point. When I_C becomes larger than I_{D14}, the current in M_{21} is zero and the start-up circuit is disabled.

5.6.3   Bipolar Core

In the bipolar front-end (Fig. 5.14), transistors Q_1 and Q_2 are biased by an array of $n + 1$ unit current sources ($n = 4$), whose current (50 nA) is derived from I_{bias}. The switches controlled by en_1 and en_2 can be configured to generate a differential output $V_{\Sigma\Delta}$ equal to either ΔV_{be} or V_{be}. If en_1 (en_2) is high and en_2 (en_1) is low, the base-emitter junction of Q_2 (Q_1) is shorted and the drain current of M_{f2} (M_{f1}) is switched off to prevent any voltage drop over the switch driven by $\overline{en_2}$ ($\overline{en_1}$). In this condition, $V_{\Sigma\Delta} = +V_{be1}$ ($V_{\Sigma\Delta} = -V_{be2}$). When both $en_{1,2}$ are high, the switches connected to the current source array are set to bias Q_1 and Q_2 either at an n:1 or at a 1:n collector current ratio, so that, respectively, either $V_{\Sigma\Delta} = \Delta V_{be}$ or $V_{\Sigma\Delta} = -\Delta V_{be}$. Because V_{be} and ΔV_{be} are not required at the same time, only two bipolar transistors are employed rather than the three used shown in Fig. 5.7.

When ΔV_{be} needs to be integrated, the accuracy of the 1:n current ratio and, hence, that of ΔV_{be} is guaranteed by a bitstream-controlled DEM scheme, which is used to swap the current sources in a way that is uncorrelated with the bitstream [4]. In successive ΔV_{be}-integration cycles, a different current source is chosen from the array to provide the unit collector current, while the other $n - 1$ sources provide the larger collector current. Mismatch errors in the current sources are thus averaged out without introducing in-band intermodulation products. Another source of error is the mismatch between Q_1 and Q_2, which can be expressed as the mismatch of their saturation currents, respectively, I_{S1} and I_{S2}. This mismatch can cause errors when generating ΔV_{be} and can be cancelled by operating the $\Sigma\Delta$ modulator in the following way. When integrating ΔV_{be}, as explained in the following section, two phases are employed: in the first phase, $Q_{1,2}$ are biased so that $I_{C1} = nI_{C2}$ and $V_{be1} - V_{be2}$ is integrated; in the second phase, $Q_{1,2}$ are biased so that $I_{C2} = nI_{C1}$ and $-(V_{be1} - V_{be2})$ is integrated. The net integrated differential charge is then

$$Q_{\Delta V_{be}} = C_a \left[\left(V_{be1}^{(1)} - V_{be2}^{(1)} \right) - \left(V_{be1}^{(2)} - V_{be2}^{(2)} \right) \right] \tag{5.22}$$

$$= C_a \frac{kT}{q} \left[\left(\ln n + \ln \frac{I_{S1}}{I_{S2}} \right) - \left(-\ln n + \ln \frac{I_{S1}}{I_{S2}} \right) \right] \tag{5.23}$$

$$= 2C_a \frac{kT}{q} \ln n \tag{5.24}$$

where the superscripts (1) and (2) refers to the voltages in the first and second phase, respectively.

To trim the sensor at room temperature, V_{be} is adjusted, as explained in Sect. 5.5.3: the collector current of Q_1 or Q_2 can be coarsely adjusted via $n - 1$ of the current sources, while the nth is driven by a digital modulator to provide a fine trim [4].

The bases of $Q_{1,2}$ are loaded by the input capacitors of the $\Sigma\Delta$. So care must be taken to ensure stable operation of the loops around $Q_{1,2}$ for any value of the collector current. Taking into consideration only one of them, the loop is comprised by three cascaded stages, Q_1, A_{f1} and M_{f1}. Miller compensation with resistive cancellation of the positive zero is introduced around A_{f1}, so that the cascade of Q_1 and A_{f1} behaves like a two-stage Miller compensated amplifier. The gain-bandwidth product can be approximated as

$$GBW \approx \frac{g_{m1}}{2\pi C_{f1}} = \frac{I_{C1}q}{2\pi kTC_{f1}} \tag{5.25}$$

where g_{m1} and I_{C1} are the transconductance and collector current of Q_1. To ensure enough phase margin for the loop, the frequency of the poles associated with A_{f1}

and M_{f1} must be larger than the worst-case GBW, i.e. that for the largest I_{C1}. A_{f1} (a current-mirror loaded differential pair) is then biased with a PTAT tail current derived from I_{bias} (equal to 400 nA at room temperature), so that its associated pole, proportional to its transconductance, moves to higher frequencies for higher temperatures, i.e. the conditions at which I_{C1} and consequently GBW are larger. The third pole due to the impedance and capacitance at the drain of M_{f1} is brought to high frequency by adding the diode-connected bipolar Q_3. The impedance of that node could have been lowered also by adding a diode-connected MOS transistor, but the use of a diode-connected bipolar is more advantageous for two reasons. Firstly, Q_3 and Q_1 form a current mirror and the collector current of Q_3 tracks I_{C1}, so that the transconductance of Q_3, and thus the third pole, are larger for a higher GBW. Secondly, for a fixed current consumption, a higher transconductance can be usually achieved by a BJT rather than with a MOS. For a fixed bias current I, this is true if

$$g_{m,BJT} = \frac{I_C}{V_t} = \frac{\beta}{\beta+1}\frac{I}{V_t} > g_{m,MOS} = \frac{I}{n_{sub}V_t} \Leftrightarrow \beta > \frac{1}{n_{sub}-1} \qquad (5.26)$$

where n_{sub} is the MOS subthreshold slope factor, $V_t = \frac{kT}{q}$ and a MOS in weak inversion has been assumed, i.e. in the operation region with highest transconductance-to-current ratio. Since n_{sub} is typically between 1.2 and 1.6 (≈ 1.5 for the devices used in this work) [26], a BJT is more efficient for $\beta > 5$.

5.6.4 Sigma-Delta ADC

A first-order $\Sigma\Delta$ modulator (Fig. 5.14) is used to sample the voltages produced by the bipolar core. The modulator implements the charge-balancing principle described in Sect. 5.4, as can be understood from the example waveforms shown in Fig. 5.15. The modulator's switched-capacitor integrator is reset at the beginning of each temperature conversion. The opamp is based on a two-stage Miller-compensated topology and achieves a minimum simulated gain of 93 dB (over process and temperature variations) with a PTAT bias current (3 μA at room temperature). Correlated Double Sampling (CDS) is used to reduce its offset and 1/f noise [27]. During phase ϕ_1, the opamp is configured as a unity-gain buffer and the signal plus offset and flicker noise are sampled on input capacitors $C_{a1,2} = 2$ pF. In the second phase ϕ_2, the offset and low frequency noise are cancelled and the charge on the input capacitors is dumped on integrating capacitors $C_{b1,2}$. Since the modulator must operate at 1.2 V, the voltage swing at the output of the integrator was scaled down by choosing $C_{b1,2} = k \cdot C_{a1,2} = 4 \cdot C_{a1,2}$. Furthermore, as shown in the timing diagram in Fig. 5.15, when $bs = 1$, only one BJT is biased and only one base-emitter voltage $-V_{be}$ is integrated, instead of the $-2V_{be}$ of previous work [4, 5]. Since a charge proportional to $2\Delta V_{be}$ is integrated when $bs = 0$ as shown in (5.24), the ratio between the charge integrated for $bs = 0$ and $bs = 1$ is equal to $-2\Delta V_{be}/V_{be}$. The factor 2 results in an equivalent factor $\alpha = 2$ in the charge-balancing conversion,

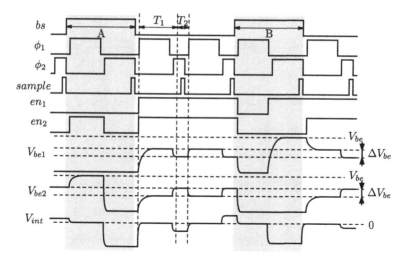

Fig. 5.15 Timing diagram and waveforms of a fragment of the temperature conversion; periods when $bs = 1$ are shown in *gray* A and B

as mentioned in Sect. 5.5.2. However, this choice means that when $bs = 1$, a V_{be}-dependent common-mode voltage will also be integrated. Imbalances in the fully differential structure of the integrator, such as mismatch in the parasitic capacitances to ground at the inverting and non-inverting input of the opamp, can result in a finite common-mode-to-differential-mode charge gain, leading to error in the output. To minimize the total integrated common-mode voltage, the sign of the input common-mode voltage is alternated in successive $bs = 1$ cycles, by setting either $V_{be1} = 0$ and $V_{be2} = V_{be}$ in ϕ_1 (period A in Fig. 5.15), or $V_{be1} = V_{be}$ and $V_{be2} = 0$ in ϕ_2 (period B).

As shown in Fig. 5.15, a longer settling time is required when one input of the modulator must switch between, say, V_{be} and $0\,\mathrm{V}$, when V_{be} is being integrated, than when one of the inputs must switch between, say, V_{be1} and V_{be2} when ΔV_{be} is being integrated. To minimize the conversion time, the length of each phase of the integrator are chosen equal either to T_1 when the input switches between 0 and V_{be} or to $T_1 < T_2$ when the input switches between V_{be1} and V_{be2}.

5.7 Temperature-Sensor Characterization

The temperature sensor (Fig. 5.16) was fabricated in a baseline TSMC 65-nm CMOS process, and was packaged in a ceramic DIL package [22]. As shown in Fig. 5.16, the active area measures $0.1\,\mathrm{mm^2}$ and it is dominated by the capacitors of the $\Sigma\Delta$'s integrator. All transistors employed in the design are thick-oxide high-threshold devices with a minimum drawn length of $0.28\,\mu\mathrm{m}$, in order to avoid any problem

Fig. 5.16 Micrograph of the
temperature sensor

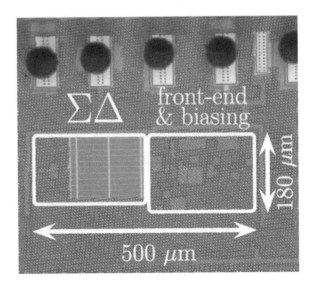

due to gate leakage, which may be significant at high temperatures. In spite of the
use of high-threshold devices, the sensor can still operate from a 1.2-V supply, from
which it draws 8.3 μA at room temperature. The supply sensitivity is 1.2°C/V at
room temperature and it is close to the required value computed in Sect. 5.3.3, which
demonstrates the low-voltage capability of the proposed NPN-based sensor. The off-
chip digital back-end decimates the output of the modulator and compensates for the
non-linearity.

With $\alpha = 2$, the modulator's bitstream average μ is limited, varying between
0.05 and 0.18 over the temperature range from -70 to 125°C. To exploit this,
a sinc^2 decimation filter was used instead of a traditional sinc filter, as this results in less quantization error over this limited range [11]. The digital non-linear
correction described in Sect. 5.5.4 has been applied off-chip, using a sixth-order
polynomial for the correction of residual non-linearities.[4] The conversion rate of
the sensor is 2.2 Sa/s (6,000 bits, $T_1 = 20\,\mu$s, $T_2 = 50\,\mu$s) at which it obtains
a quantization-noise-limited resolution of 0.03°C. A set of devices was measured
over the temperature range from -70 to 125°C. After digital compensation for
systematic non-linearity, the inaccuracy (Fig. 5.17) was 0.5°C (3σ, 12 devices). This
improved to 0.2°C (3σ, 16 devices) after trimming at 30°C (Fig. 5.18). Since with
those settings the noise level of the sensor is limited by the quantization noise, it
is possible (in a redesign) to increase the thermal noise level without affecting the
performance. This would allow for smaller capacitances in the $\Sigma\Delta$ ($C_{a1,2}$ and $C_{b1,2}$
in Fig. 5.14), resulting in less area and a faster conversion.

---

[4]A third or fourth-order polynomial would be sufficient to compensate for the third-order non-
linearity of V_{be}, which is usually the main residual non-linearity. A sixth-order polynomial is
employed in this work to compensate the strong non-linearity at high temperature due to leakage.

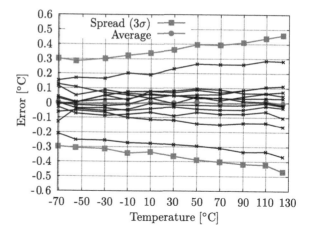

Fig. 5.17 Measured temperature error (with $\pm 3\sigma$ limits) of 12 samples after batch calibration

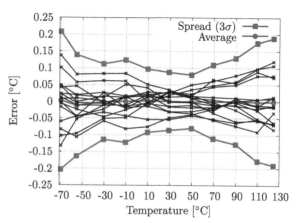

Fig. 5.18 Measured temperature error (with $\pm 3\sigma$ limits) of 16 samples after trimming at 30°C

A summary of the sensor's performance and a comparison to the state-of-the-art for CMOS temperature sensors is reported in Table 5.1. The sensor's untrimmed accuracy is ten times better than previous designs in deep-submicron CMOS and both its batch-calibrated and trimmed accuracy are comparable with sensors realized in larger-feature-size processes. These advances are enabled by the use of vertical NPN transistors as sensing elements, the use of precision circuit techniques, such as DEM and dynamic offset compensation, and a single room-temperature trim. In particular, the use of NPNs, rather than the PNPs of previous work, enables sensing at much lower temperature (−70°C) while operating from a low supply voltage (1.2 V). Such NPNs can be made without process modifications by exploiting the availability of deep n-well diffusions in most deep-submicron CMOS processes. Thus, accurate temperature sensors can still be designed in advanced deep-submicron CMOS processes, despite the limited current-gain of the available bipolar transistors and the need for operation at low supply voltages.

Table 5.1 Performance summary of the temperature sensor and comparison with previously published CMOS temperature sensors

	This work	[7]	[12]	[5]
Technology	65 nm CMOS	32 nm CMOS	0.16 μm CMOS	0.7 μm CMOS
Chip area	0.1 mm^2	0.02 mm^2	0.26 mm^2	4.5 mm^2
Supply current	8.3 μA	1.5 mA	6 μA	25 μA
Supply voltage	1.2–1.3 V	1.05 V	1.8 V	2.5–5.5 V
Supply sensitivity	1.2°C/V	N.A.	0.2°C/V	0.05°C/V
Output rate	2.2 Sa/s	1 kSa/s	10 Sa/s	10 Sa/s
Energy per conversion	4.5 μJ	1.6 μJ	0.9 μJ	12.5 μJ
Resolution	0.03°C^a	0.15°C (1σ)	0.018°C (1σ)	0.025°C (1σ)
Temperature range	−70 to 125°C	−10 to 110°C	−40 to 125°C	−55 to 125°C
Inaccuracy (untrimmed)	0.5°C (3σ)	<5°C	0.5°C (3σ)	0.25°C (3σ)
Inaccuracy (trimmed)	0.2°C (3σ)	N.A.	0.25°C (3σ)	0.1°C (3σ)

^{a}The resolution is limited by quantization noise and hence 1 LSB is reported

5.8   Temperature Compensation of the Time Reference

To investigate the level of accuracy reachable by the mobility-based time reference, two distinct temperature compensation schemes have been implemented, requiring either a single-point trim or a two-point trim [28]. For both schemes, the resolution of the divider ratio N_{div} has been kept to 13 bits. As discussed in Sect. 5.2, even a resolution of approximately 10 bits is high enough when aiming for accuracy in the order of 1%. However, a higher resolution is preferred to show that inaccuracy even lower than 1% can be reached.

For a single-point trim, the oscillation frequency of each sample at the trim temperature $f_{osc}(T_{trim})$ is measured. A seventh-order polynomial $P_7(\cdot)$, whose coefficients are fixed for all the samples, is then obtained via a batch calibration. The divider factor N_{div} is computed as

$$N_{div} = \frac{f_{osc}(T_{trim})}{f_{nom}} P_7(\mu) \qquad (5.27)$$

where $f_{nom} = 150\,\text{kHz}$ is the nominal frequency of oscillation, i.e. the desired output frequency.

For a two-point trim, the following procedure is adopted. The oscillator frequency f_{osc} and the on-chip temperature-sensor decimated output μ are measured at two different temperatures, $T_{trim,1}$ and $T_{trim,2}$. Those data are used to interpolate the frequency using the interpolant

$$f_{osc} = A \cdot \mu_{PTAT}^B \qquad (5.28)$$

where μ_{PTAT} is computed from the temperature-sensor output using (5.17) and A and B are the trim parameters for each sample. A fourth-order polynomial $Q_4(\cdot)$ is obtained from batch calibration so that the divider factor N computed for each sample is

$$N_{div} = \frac{1}{f_{nom}} A \cdot \{\mu_{PTAT}\,[Q_4(\mu)]\}^B \qquad (5.29)$$

The polynomial[5] $Q_4(\cdot)$ is required to compensate for the fact that the power-law interpolant in (5.28) only approximately describes the temperature dependence of the electron mobility, especially over a wide temperature range.

---

[5]Note that the order of the polynomials $P_7(\cdot)$ and $Q_4(\cdot)$ is the minimum required for the error due to the non-linearity of the compensation to be negligible compared to the spread among the samples.

5.9 Experimental Results

The time reference was fabricated in a standard 65-nm CMOS process (Fig. 5.19), by integrating on the same die the oscillator and the temperature sensor presented, respectively, in Chap. 4 and in the previous part of this chapter [28]. The circuit occupies $0.2\,mm^2$ ($0.1\,mm^2$ for the oscillator and $0.1\,mm^2$ for the temperature sensor) and uses only 2.5-V I/O thick-oxide MOS devices . For flexibility, some control logic, the temperature sensor's $sinc^2$ decimation filter (employing 6,000 bitstream samples to produce one temperature reading) and the reference voltages (V_R, V_{r1}, V_{r2}) were implemented off-chip. The reference draws $42.6\,\mu A$ ($34.3\,\mu A$ for the oscillator and $8.3\,\mu A$ for the temperature sensor) from a 1.2-V supply at room temperature. Using (5.6), the worst-case supply sensitivity is estimated to be 1.2%/V ($\frac{\Delta f_{osc}}{f_{osc}}/\Delta V_{dd} = 0.6\%/V$, $\Delta T/\Delta V_{dd} = 1.2°C/V$). In order to flexibly test different compensation schemes, the temperature compensation scheme has been implemented with an off-chip FPGA, as shown in Fig. 5.20. However, if the temperature compensation had been implemented on-chip, its consumption would have been negligible. From simulations, the power consumption of a 13-bit frequency divider implemented in 65-nm CMOS is below $0.3\,\mu W$ at room temperature for an input frequency of 150 kHz and a 1.2-V supply. The non-linear-mapping block's consumption would also be low considering that the value of N_{div} must be updated only at the output rate of the temperature sensor, which is in the order[6] of 1 Hz.

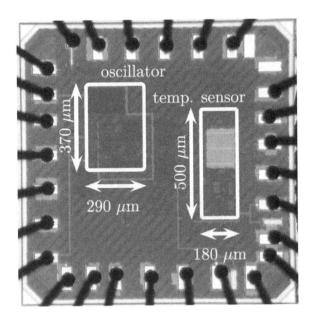

Fig. 5.19 Die micrograph of the test chip

---

[6]The up-date rate of N_{div} depends on the temperature variations rate in the chosen application.

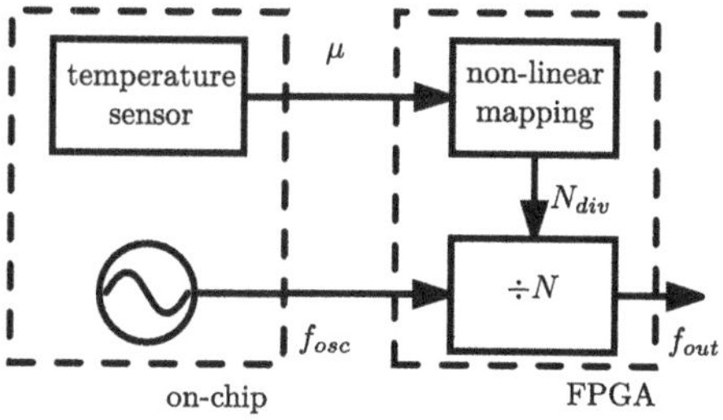

Fig. 5.20 System block diagram of the mobility-based time reference showing the partition in the implementation between on-chip circuitry and FPGA

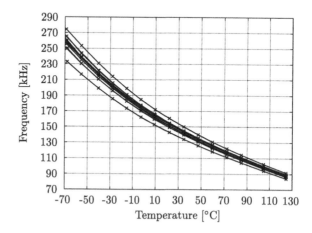

Fig. 5.21 Uncompensated oscillator output frequency (f_{osc})

Measurements on 12 samples from one batch were performed over the temperature range from -70 to $+125°C$ using a temperature-controlled oven. For those measurements, the reference voltages were set to $V_R = 0.25\,V$, $V_{r1} = 1.6\,V$ and $V_{r2} = 1.2\,V$ and the supply of the comparator of Fig. 4.6 has been raised to 1.5 V. This ensures that MOS capacitor C_A and C_B are biased in deep inversion to prevent the spread of their threshold voltage from affecting the spread of the reference, and thus masking the inherent accuracy of the mobility-based reference. The temperature of the samples was measured with a Pt100 platinum thermometer and compared to the temperature reading of the on-chip temperature sensor. The temperature sensor was not trimmed, since, as demonstrated by the experimental results presented below, its untrimmed inaccuracy of $0.5°C$ (3σ) over the range from -70 to $+125°C$ is low enough not to significantly affect the reference's accuracy.

Figure 5.21 shows the uncompensated output frequency of the oscillator. At room temperature, its maximum deviation from the average is $\pm6\%$. The temperature compensation has then been implemented off-line in Matlab. First, the samples were trimmed at $T_{trim} = 22°C$ and compensated with an external Pt100 and an ideal temperature compensation curve. In those conditions, the maximum error is $\pm2.6\%$ over the military range from -55 to $125°C$. Then, the compensation polynomial $P_7(\cdot)$ (see Sect. 5.8) was extracted from batch calibration of the 12 devices. After a single-point trim at $T_{trim} = 22°C$, the error when compensating with the

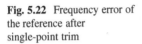

Fig. 5.22 Frequency error of
the reference after
single-point trim

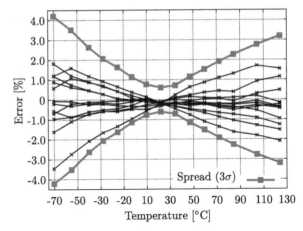

Fig. 5.23 Frequency error of
the reference after two-point
trim

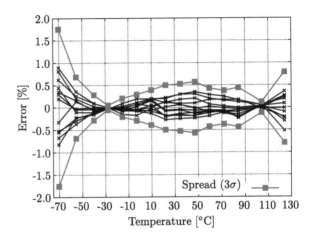

on-chip temperature sensor is less than $\pm 2.7\%$ (Fig. 5.22). Finally, a two-point trim
at $T_{trim,1} = -27°C$ and $T_{trim,2} = 105°C$ was employed and the error improved
to $\pm 0.5\%$ using another compensating polynomial $Q_4(\cdot)$ (see Sect. 5.8) extracted
from a batch calibration of the 12 devices (Fig. 5.23). The temperature of the two
trimming points has been chosen to optimize the accuracy of the time reference
over the temperature range of interest. For the adopted compensation schemes, the
resolution of the integer divider factor N_{div} in Fig. 5.1 has been limited to 13 bits.

In order to study the dynamic performance of the temperature compensation, one
of the samples already tested in the temperature-controlled oven has been tested in
the presence of rapid temperature variations. Figure 5.24 shows the static frequency
error of the tested sample over the temperature range, i.e. the frequency error under
static temperature conditions. Note that the selected sample is one of those with
the least static error, so that any dynamic error can be easily observed. In this case,
real-time temperature compensation has been performed on the FPGA, while the

Fig. 5.24 Frequency error of
the sample used for
time-domain measurements
(after a single-point trim at
room temperature)

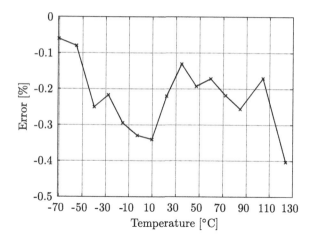

sample (packaged in a ceramic DIL28 package) was heated by a flux of hot air. For
those measurements, the reference voltages were set to $V_R = 0.2\,\text{V}$, $V_{r1} = 1.2\,\text{V}$
and $V_{r2} = 0.8\,\text{V}$, $f_{osc} = 96\,\text{kHz}$ and all supplies were set to 1.2 V to simulate the
operation of the system while powered by a 1.2-V battery. The temperature sensor
was sampled[7] at $f_{sample} = 3\,\text{Sa/s}$. The $\Sigma\Delta$ modulator in the temperature sensor is
operated as an incremental converter: it is first reset and a temperature reading
is produced after $1/f_{sample}$. In the next period $1/f_{sample}$, such temperature reading
is produced to compute N_{div} and compensate f_{out}. The output frequency is then
affected by errors due to the delay of $1/f_{sample}$ in the temperature compensation.
Figure 5.25 shows the measured time-domain waveforms acquired by the FPGA,
including the temperature reading of the on-chip temperature sensor, the period of
f_{out} and the divider ratio N_{div}. When hot air is applied, the on-chip temperature rises
from room temperature to approximately 100°C with an exponential response with
time constant $\tau_1 = 20\,\text{s}$. Due to the fast temperature variation, a jump is observed
in the oscillation period due to the temperature compensation delay. The amplitude
of the jump can be computed by considering the temperature reading error ΔT due
to the delay and by applying (5.5). The amplitude of the jump is given by

$$\Delta t_{jump} = t_0 \left(\frac{1}{f_{osc}} \frac{\partial f_{osc}}{\partial T} \right) \Delta T \approx t_0 \left(\frac{1}{f_{osc}} \frac{\partial f_{osc}}{\partial T} \right) \left(-\frac{T_{jump}}{\tau_1} \frac{1}{f_{sample}} \right) \qquad (5.30)$$

where t_0 is the oscillation period of the output signal ($1/f_{osc}$) in the steady state, ΔT
is the equivalent temperature error due to the compensation delay and $T_{jump} = 75°C$

---

[7]Note that a faster sampling rate is adopted for the temperature sensor with respect to that
(2.2 Sa/s) employed for the measurements presented in Sect. 5.7. Though this could result in a
larger temperature error, it is allowed because, as shown earlier in this section, the full accuracy of
the temperature sensor is not required for the reference's compensation.

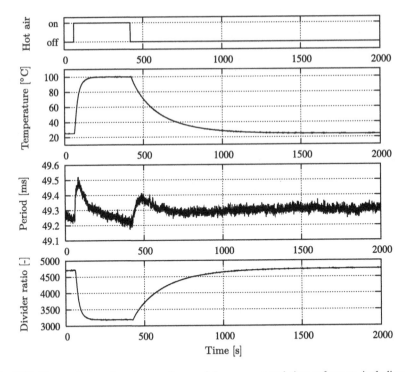

Fig. 5.25 Measured time-domain waveforms of the compensated time reference, including the on-chip temperature-sensor reading, the period of the output signal $(1/f_{out})$ and the divider ratio N_{div}, when the sample is heated by a flux of hot air (whose timing is shown on the *upper plot*)

is the jump in the die temperature. ΔT has been calculated by considering an exponential settling of the temperature with time constant τ_1. Using (5.5) and the data from Sect. 5.3, $\Delta t_{jump} \approx 0.3$ ms, which is in good agreement with the value measured in Fig. 5.25. When the flow of hot air stops, the temperature drops exponentially to the room temperature.[8] Even if a negative jump in the oscillation period is expected at the onset of the cooling, it is not visible in the measurements. This is because the cooling transient is slow enough to make the dynamic error much smaller than the static error reported in Fig. 5.24. Such static error also accounts for the deviations of the oscillation period during the cooling. The oscillation period is also affected by random noise with standard deviation of 0.1%, in agreement with the long-term jitter measurements shown previously in Chap. 4. This proves that also under excitation with large and fast temperature variations, the time reference error is kept in the order of 1%.

---

[8]The time constant of the exponential settling is in this case $\tau_2 = 180$ s due to the lack of induced air flow on the sample and the consequent increase in thermal resistance R_{pkg}.

Table 5.2 Performance summary and comparison

Reference	[29]	[30]	[31]	This work	
Frequency	6 MHz	10 MHz	30 MHz	150 kHz	
Supply	1.2 V	1.2 V	3.3 V	1.2 V	
Power	66 μW	80 μW	180 μW	51 μW	
Technology	65 nm	0.18 μm	0.35 μm	65 nm	
Temp. range (°C)	0 ∼ 120	−20 ∼ 100	−20 ∼ 100	−55 ∼125	
Inaccuracy (%)	±0.9	±0.4	±0.7	±0.5	±2.7
Calibration Samples	Single	N.A.[a]	N.A.[a]	Double	Single
tested over temp.	4	1	1	12	

[a]No calibration defined for a single sample

The time reference's performance is summarized in Table 5.2 and is compared to other low-power fully integrated CMOS time references. These references are a subset of those presented in Sect. 3.9 and are characterized by a power consumption of the order of 100 μW or lower. The silicon area and the power consumption of the proposed time reference are comparable to the other designs in Table 5.2, even if a fair comparison is challenging due to the lower oscillator frequency chosen in this work. Because of the temperature compensation by frequency division shown in Fig. 5.20, the output frequency is only 20 Hz. However, both in this work and in those cited in Table 5.2, the power dissipated for temperature compensation is, to first order, independent of the output frequency. With regard to the performance of the temperature compensation scheme, the proposed time reference achieves an accuracy comparable to the state-of-the-art over a wider temperature range and for significantly more measured samples.

5.10 Conclusions

In Chap. 4, it was shown that an accurate temperature compensation scheme must be implemented to fully exploit the accuracy of the mobility-based oscillator. Such a compensation scheme requires two main components: a sensor to provide an accurate temperature reading and a tuning mechanism that, based on the output of the sensor, corrects the oscillator frequency, or generates a derivative frequency, to obtain a temperature-independent frequency. The error introduced by each of those components should be much smaller than the output spread of the mobility-based oscillator of Chap. 4, which is of the order of 1% over the commercial temperature range. Consequently, the inaccuracy of the reference due to error in temperature sensing and in frequency compensation must be of the order of 0.1%.

The frequency compensation can be made very accurate by adopting a digital compensation scheme. The oscillator output can be multiplied or divided by a temperature-dependent integer factor to obtain, respectively, a temperature-independent RF reference and a temperature-independent very-low-frequency reference. Those signals are needed in the WSN node described in Chap. 2 to enable RF communication in the right channel of the available spectrum and to enable timing synchronization between different node in the WSN. Since the RF and low-frequency reference must be present in the node in any case, the circuitry can be re-used for the temperature compensation, without any hardware or power overhead. It has been shown that, taking into consideration the requirements for the WSN node, both signals can be generated by allowing a 10-bit resolution for the temperature compensation, i.e. an inaccuracy of less than 0.1%, if the mobility-based oscillator frequency is of the order of 100 kHz. Thus, the oscillator from Chap. 4 has been employed for demonstrating the temperature compensation.

The other error source for the temperature compensation is the inaccuracy of the temperature sensor. To make sure that this block does not contribute more than 0.1% to the total inaccuracy of the reference, the inaccuracy of the sensor must be below 0.2°C. This level of accuracy is not easily achievable by temperature sensors integrated in a deep-submicron technology. Since such cutting-edge technology must be adopted for the whole WSN node, as explained in Chap. 2, specific techniques to design a highly accurate deep-submicron temperature sensor have been introduced and analyzed. The use of vertical NPN transistors as sensing elements, the use of precision circuit techniques, such as DEM and dynamic offset compensation, and a single room-temperature trim, have made possible the implementation of a low-voltage sensor with untrimmed accuracy 10× better than previous designs in deep-submicron CMOS and accuracy comparable with sensors realized in larger-feature-size processes. The presented sensor achieves a measured inaccuracy after a single point-trim of 0.2°C (3σ) over the temperature range from −70 to 125°C.

In order to experimentally show the validity of the proposed temperature compensation scheme, the mobility-based oscillator and the temperature sensor were integrated on the same die and the real-time frequency compensation has been implemented off-chip on an FPGA. The integrated temperature-compensated time reference achieves an inaccuracy less than ±2.7% after single-point trim and less than ±0.5% after two-point trim over the military temperature range. This demonstrates that time references with inaccuracies less than 1% over a wide temperature range can be realized with MOS transistors, even in nanometer CMOS. Those references are accurate enough for WSN applications, while working at low-voltage and low-power, as required for use in autonomous sensor nodes.

References

1. NXP Semiconductor (2000) Integrated circuit packages data handbook. http://www.standardics.nxp.com/packaging/handbook/ Accessed May 2012
2. Linear Technologies (2006) Thermal resistance table. http://www.linear.com/designtools/packaging/Linear_Technology_Thermal_Resistance_Table.pdf Accessed December 2010
3. Sofia J (1995) Analysis of thermal transient data with synthesized dynamic models for semiconductor devices. IEEE Trans Comp Packag Manuf Technol A 18(1):39–47. DOI 10.1109/95.370733

4. Pertijs M, Makinwa K, Huijsing J (2005) A CMOS smart temperature sensor with a 3σ inaccuracy of ±0.1 °C from -55°C to 125°C. IEEE J Solid State Circ 40(12):2805–2815. DOI 10.1109/JSSC.2005.858476

5. Aita A, Pertijs M, Makinwa K, Huijsing J (2009) A CMOS smart temperature sensor with a batch-calibrated inaccuracy of ±0.25°C (3σ) from -70°C to 130°C. In: ISSCC Dig. Tech. Papers, pp 342–343, 343a. DOI 10.1109/ISSCC.2009.4977448

6. Duarte D, Geannopoulos G, Mughal U, Wong K, Taylor G (2007) Temperature sensor design in a high volume manufacturing 65nm CMOS digital process. In: Proc. IEEE Custom Integrated Circuits Conf. (CICC), pp 221–224. DOI 10.1109/CICC.2007.4405718

7. Lakdawala H, Li Y, Raychowdhury A, Taylor G, Soumyanath K (2009) A 1.05 V 1.6 mW 0.45 °C 3σ-resolution $\Sigma\Delta$-based temperature sensor with parasitic-resistance compensation in 32 nm CMOS. IEEE J Solid State Circ (12):3621–3630

8. Floyd MS, Ghiasi S, Keller TW, Rajamani K, Rawson FL, Rudbio JC, Ware MS (2007) System power management support in the IBM POWER6 microprocessor. IBM J Res Develop 51(6):733–746

9. Saneyoshi E, Nose K, Kajita M, Mizuno M (2008) A 1.1V 35 μm × 35 μm thermal sensor with supply voltage sensitivity of 2°C/10% -supply for thermal management on the SX-9 supercomputer. In: IEEE Symposium on VLSI Circuits Dig. Tech. Papers, pp 152–153. DOI 10.1109/VLSIC.2008.4585987

10. Chen P, Chen CC, Peng YH, Wang KM, Wang YS (2010) A time-domain SAR smart temperature sensor with curvature compensation and a 3σ inaccuracy of -0.4°C ~ $+0.6$°C over a 0°C to 90°C range. IEEE J Solid State Circ 45(3):600–609. DOI 10.1109/JSSC.2010.2040658

11. Pertijs MAP, Huijsing JH (2006) Precision temperature sensors in CMOS technology. Springer, Dordrecht

12. Souri K, Kashmiri M, Makinwa K (2009) A CMOS temperature sensor with an energy-efficient Zoom ADC and an inaccuracy of ±0.25°C (3σ) from -40°C to 125°C. In: ISSCC Dig. Tech. Papers, pp 310–311, 311a. DOI 10.1109/ISSCC.2009.4977448

13. Krummenacher P, Oguey H (1989) Smart temperature sensor in CMOS technology. Sensor Actuator A Phys 22(1–3):636–638. DOI 10.1016/0924-4247(89)80048-2

14. Kim JP, Yang W, Tan HY (2003) A low-power 256-Mb SDRAM with an on-chip thermometer and biased reference line sensing scheme. IEEE J Solid State Circ 38(2):329–337. DOI 10.1109/JSSC.2002.807170

15. Szajda K, Sodini C, Bowman H (1996) A low noise, high resolution silicon temperature sensor. IEEE J Solid State Circ 31(9):1308–1313. DOI 10.1109/4.535415

16. Creemer J, Fruett F, Meijer G, French P (2001) The piezojunction effect in silicon sensors and circuits and its relation to piezoresistance. IEEE Sensor J 1(2):98–108. DOI 10.1109/JSEN.2001.936927

17. Meijer G, Gelder RV, Nooder V, Drecht JV, Kerkvliet H (1989) A three-terminal intergrated temperature transducer with microcomputer interfacing. Sensor Actuator 18(2):195–206. DOI 10.1016/0250-6874(89)87018-0

18. Kashmiri SM, Pertijs MAP, Makinwa KAA (2010) A thermal-diffusivity-based frequency reference in standard CMOS with an absolute inaccuracy of ±0.1% from -55°C to 125°C. IEEE J Solid State Circ 45(12):2510–2520

19. Norsworthy S, Schreier R, Temes G (eds) (1996) Delta-sigma data converters: theory, design, and simulation. Wiley, Hoboken

20. Meijer G, Wang G, Fruett F (2001) Temperature sensors and voltage references implemented in CMOS technology. IEEE Sensor J 1(3):225–234. DOI 10.1109/JSEN.2001.954835

21. Meijer GC (1986) Thermal sensors based on transistors. Sensors Actuators 10(1–2):103–125. DOI 10.1016/0250-6874(86)80037-3

22. Sebastiano F, Breems L, Makinwa K, Drago S, Leenaerts D, Nauta B (2010a) A 1.2-V 10-μW NPN-based temperature sensor in 65-nm CMOS with an inaccuracy of 0.2°C (3σ) from -70°C to 125°C. IEEE J Solid State Circ 45(12):2591–2601

23. Aita A, Makinwa K (2007) Low-power operation of a precision CMOS temperature sensor based on substrate PNPs. In: IEEE Sensors, pp 856–859. DOI 10.1109/ICSENS.2007.4388536

24. You F, Embabi H, Duque-Carrillo J, Sanchez-Sinencio E (1997) An improved tail current source for low voltage applications. IEEE J Solid State Circ 32(8):1173–1180. DOI 10.1109/4.604073

25. Baker RJ, Li HW, Boyce DE (1997) CMOS circuit design, layout, and simulations. IEEE, New York, p 637

26. Pouydebasque A, Charbuillet C, Gwoziecki R, Skotnicki T (2007) Refinement of the subthreshold slope modeling for advanced bulk CMOS devices. IEEE Trans Electron Dev 54(10):2723–2729. DOI 10.1109/TED.2007.904483

27. Enz C, Temes G (1996) Circuit techniques for reducing the effects of op-amp imperfections: autozeroing, correlated double sampling, and chopper stabilization. Proc IEEE 84(11):1584–1614

28. Sebastiano F, Breems L, Makinwa K, Drago S, Leenaerts D, Nauta B (2011) A 65-nm CMOS temperature-compensated mobility-based frequency reference for wireless sensor networks. IEEE J Solid State Circ, 46(7):1544–1552

29. De Smedt V, De Wit P, Vereecken W, Steyaert M (2009) A 66 μW 86 ppm/$^\circ$C fully-integrated 6 MHz wienbridge oscillator with a 172 dB phase noise FOM. IEEE J Solid State Circ 44(7):1990–2001. DOI 10.1109/JSSC.2009.2021914

30. Lee J, Cho S (2009) A 10MHz 80μW 67 ppm/$^\circ$C CMOS reference clock oscillator with a temperature compensated feedback loop in 0.18μm CMOS. In: 2009 Symposium on VLSI Circuits Dig. Tech. Papers, pp 226–227

31. Ueno K, Asai T, Amemiya Y (2009) A 10MHz 80μW 67 ppm/$^\circ$C CMOS reference clock oscillator with a temperature compensated feedback loop in 0.18μm CMOS. In: Proc. ESSCIRC, pp 226–227

Chapter 6
Conclusions

This final chapter presents a summary of the main findings of this work and a short overview of the possible applications and research topics originating from this work that could be investigated in the future.

6.1 Main Findings

- Adopting a duty-cycled wake-up radio and an impulse-based modulation scheme in a Wireless Sensor Network (WSN) can lower the accuracy requirement of the on-board time reference down to 1% (Chap. 2).
- The inaccuracy of mobility-based time references depends both on the device used and on the selected packages: in a 0.16-μm-CMOS process, inaccuracies as low as 1% over the military temperature range can be achieved using thin-oxide transistor and ceramic packages and can be kept below 2% over the same temperature range when using low-cost plastic packages (Chap. 4).
- Vertical NPN transistors based on deep n-well diffusions are the preferred devices for the implementation of temperature sensors in deep-submicron CMOS processes (Chap. 5).
- By using vertical NPN transistors as sensing elements, precision circuit techniques, such as Dynamic Element Matching (DEM) and dynamic offset compensation, and a single room-temperature trim, temperature sensors in deep-submicron CMOS processes can achieve a 3σ inaccuracy of $\pm0.2°$C over the temperature range from -70 to $125°$C (Chap. 5).
- The inaccuracy of temperature-compensated mobility-based time references implemented in a deep-submicron process and packaged in ceramic can be as low as $\pm2.7\%$ over the military temperature range, after a room-temperature trim, and as low as $\pm0.5\%$, after a two-point trim, making them suitable for WSN applications (Chap. 5).

F. Sebastiano et al., *Mobility-based Time References for Wireless Sensor Networks*, Analog Circuits and Signal Processing, DOI 10.1007/978-1-4614-3483-2_6, © Springer Science+Business Media New York 2013

6.2 Applications

The mobility-based temperature-compensated time reference has been conceived to enable the implementation of a fully integrated node for WSNs. The presented 65-nm CMOS time reference could be integrated together with the RF front-end described in [1], which has been developed in the same technology and complying with the impulse-radio modulation scheme described in this work. By even accounting for additional digital circuitry occupying $0.5\,mm^2$, the resulting wireless sensor node would take a silicon area smaller than $1\,mm^2$. To initiate the communication with a neighboring node, only an antenna and a battery should be plugged externally. Moreover, a field trial could be readily set up by using the reading of the on-board temperature sensor as the environmental parameter to be monitored. This would facilitate the development of WSNs consisting of a huge number of extremely small, cheap and energy-autonomous nodes.

Additionally, the individual circuit blocks developed in this work can serve different applications. For example, the time reference can be employed as a real-time clock embedded in larger Systems-on-Chip (SoCs), such as microcontrollers, and the deep-submicron temperature sensor can be used for the thermal management of microprocessors [2, 3] or for the compensation of frequency references based on different physical principles.

6.3 Future Research

- Mismatch of circuit components is one of the sources of inaccuracy in the proposed design of the mobility-based reference. Its effect is attenuated by a single-point trim and the use of devices with large area, but dynamic mismatch compensation techniques, such as DEM and chopping used in the temperature sensor described in Chap. 5, could also be adopted, with a possible improvement in accuracy and decrease of silicon area.
- The use of a relaxation oscillator in the mobility-based reference requires the design of a very fast comparator, which consumes a substantial fraction of the whole power. The design of such block is even more challenging when using more mature and slower processes. Possible alternatives would be a ring oscillator whose delay elements are referenced to the mobility or a locked oscillator referenced to the charge of a capacitor with a mobility-dependent current. This could be achieved by proper biasing (by current or voltage) of the elements. While a lower inaccuracy can be expected, the power consumption can be lower and no fast switching circuits, as the autolatch comparator in Chap. 4, would be required.
- RC-based references can be investigated using an approach analogous to the one followed in this work. The temperature coefficient of integrated resistors (preferably polysilicon resistors for their versatility and their low temperature

coefficient) could be compensated by an integrated temperature sensor. An advantage would be that the temperature sensor's inaccuracy could be relaxed due to the lower temperature coefficient of resistors compared to that of mobility. To assess the expected inaccuracy, the spread of the temperature coefficient of integrated resistors should be evaluated.

- The architecture of the mobility-based reference can be simplified if a dedicated temperature sensor is avoided. A simpler temperature compensation could be operated by compensating the temperature coefficient of the mobility with a physical parameter with complementary temperature dependence (yet to be identified), as is usually done in RC oscillators, in which resistors with positive and negative temperature coefficients are combined in series to get an almost temperature-independent resistance.

- The jitter of the mobility-based oscillator is averaged over multiple cycles when it is used as time reference to measure intervals in the order of tens of milliseconds. When the reference frequency is multiplied by the PLL briefly described in Chap. 5, however, the phase noise will be amplified by the multiplication factor of the PLL. The phase noise of the reference could be reduced by using a narrow-band loop filter in the PLL (preferably implemented in the digital domain) or by letting the RF oscillator free running and only periodically calibrating it frequency.

References

1. Drago S, Leenaerts D, Sebastiano F, Breems L, Makinwa K, Nauta B (2010) A 2.4 GHz 830 pJ/bit duty-cycled wake-up receiver with −82 dbm sensitivity for crystal-less wireless sensor nodes. In: ISSCC Dig. of Tech. Papers, pp 224–225
2. Floyd MS, Ghiasi S, Keller TW, Rajamani K, Rawson FL, Rudbio JC, Ware MS (2007) System power management support in the IBM POWER6 microprocessor. IBM J Res Develop 51(6):733–746
3. Poirier C, McGowen R, Bostak C, Naffziger S (2005) Power and temperature control on a 90nm Itanium®-family processor. In: ISSCC Dig. Tech. Papers, vol 1, pp 304–305. DOI 10.1109/ ISSCC.2005.1493990

Appendix A
Derivation of the Accuracy of Mobility-Based Oscillator

A.1 Composite MOS Transistor

A composite MOS transistor M_x is composed by the parallel combination (short-circuited drain, gate and source) of n unit transistors $M_{x,1}$, $M_{x,2}$, ... $M_{x,n}$. Its drain current can be expressed as

$$I = \sum_{i=1}^{n} \frac{\beta_{x,i}}{2} (V_{Gsx} - V_{tx,i})^2 \tag{A.1}$$

where $\beta_{x,i}$ and $V_{tx,i}$ are the current factor β and threshold voltage of $M_{x,i}$, respectively. The following notation is adopted

$$\Delta I_x = I_x - I_{x,0} \quad I_{x,0} = \langle I \rangle \tag{A.2}$$

$$\Delta V_{tx,i} = V_{tx,i} - V_{tx} \quad V_{tx} = \langle V_{tx,i} \rangle \tag{A.3}$$

$$\Delta \beta_{x,i} = \beta - \beta_u \quad \beta_u = \langle \beta_{x,i} \rangle \quad \beta_{x,0} = n\beta_u \tag{A.4}$$

Note that in the previous definitions and in the rest of the appendix, the average is intended over the set of the implementations of M_x. Using a first-order Taylor expansion and neglecting higher-order terms, we can derive

$$\Delta I_x \approx \sum_{i=0}^{n} \frac{\partial I_x}{\beta_{x,i}} \Delta \beta_{x,i} + \frac{\partial I_x}{V_{tx,i}} \Delta V_{tx,i} \tag{A.5}$$

$$= \sum_{i=0}^{n} \frac{\Delta \beta_{x,i}}{2} (V_{Gsx} - V_{tx,i})^2 - \beta_{x,i} (V_{Gsx} - V_{tx,i}) \Delta V_{tx,i} \tag{A.6}$$

F. Sebastiano et al., *Mobility-based Time References for Wireless Sensor Networks*,
Analog Circuits and Signal Processing, DOI 10.1007/978-1-4614-3483-2,
© Springer Science+Business Media New York 2013

$$\approx \sum_{i=0}^{n} \frac{\Delta \beta_{x,i}}{2} (V_{Gsx} - V_{tx0})^2 - \beta_u (V_{Gsx} - V_{tx0}) \Delta V_{tx,i} \qquad (A.7)$$

$$\approx \frac{\sum_{i=0}^{n} \Delta \beta_{x,i}}{2} (V_{Gsx} - V_{tx0})^2 +$$

$$-\beta_{x,0} (V_{Gsx} - V_{tx0}) \frac{\sum_{i=0}^{n} \Delta V_{tx,i}}{n} \qquad (A.8)$$

Defining

$$I' = \frac{\sum_{i=0}^{n} \Delta \beta_{x,i}}{2} \left(V_{Gsx} - \frac{\sum_{i=0}^{n} \Delta V_{tx,i}}{n} \right)^2 \qquad (A.9)$$

$$\Delta I'_x = I'_x - \langle I_x \rangle \qquad (A.10)$$

it is possible to find for $\Delta I'_x$ a similar expression as done above for ΔI_x and prove that $\Delta I_x \approx \Delta I'_x$.

The previous calculation show that for small deviation from the average value, the current of a composite transistor can be expressed as

$$I_x = \frac{\beta_x}{2} (V_{Gsx} - V_{tx})^2 \qquad (A.11)$$

where

$$\beta_x = \beta_{x,0} + \Delta \beta_x \quad \Delta \beta_x = \sum_{i=0}^{n} \Delta \beta_{x,i} \qquad (A.12)$$

$$V_{tx} = V_{tx,0} + \Delta V_{tx} \quad \Delta V_{tx} = \frac{1}{n} \sum_{i=0}^{n} \Delta V_{tx,i} \qquad (A.13)$$

The standard deviation of the parameters defined above are simply given by

$$\sigma_{\frac{\Delta \beta_x}{\beta_x}} = \frac{1}{\sqrt{2n}} \sigma_{\frac{\Delta \beta_u}{\beta_u}} \qquad (A.14)$$

$$\sigma_{\Delta V_{tx}} = \frac{1}{\sqrt{2n}} \sigma_{\Delta V_{tx,u}} \qquad (A.15)$$

where $\sigma_{\frac{\Delta \beta_u}{\beta_u}}$ and $\sigma_{\Delta V_{tx,u}}$ are the standard deviation of the relative β *mismatch* and threshold voltage *mismatch* between *two* unit transistors, which are the data usually available in design rule manuals. A factor $\sqrt{2}$ appears in the last two equations because the error has been previously defined as the difference between one transistor and the *average* transistor.

A.2   Accuracy of the Mobility-Based Current Reference

A.2.1   Effect of Mismatch and Finite Output Resistance

In the following, an expression for the current I_1 in the circuit of Fig. A.1 is provided, considering the effect of mismatch and transistor finite output resistance. It is assumed that transistor M_1 and M_3 are composite transistors (see Sect. A.1) composed by m_3 and m_4 unit transistors, respectively, where

$$m = \frac{m_3}{m_1} \tag{A.16}$$

The drain currents of M_1 and M_3 are given by

$$I_{1,3} = \frac{\beta_{1,3}}{2} (V_{GS1,2} - V_{t1,2})^2 (1 + \lambda_{1,3} V_{DS1,3}) \tag{A.17}$$

where $\lambda_{1,3} = 1/(I_{1,3}|_{V_{DS1,3}=0} \, r_{o1,3})$ is the channel-length-modulation factor and $r_{o1,3}$ is the output resistance (between drain and source) of the transistors.

From inspection of the circuit, the following system of equation can be written

$$\begin{cases} I_3 = nI_1 \\ V_{GS3} = V_{GS1} + V_R \end{cases} \tag{A.18}$$

By using (A.17) and solving the system, we obtain

$$I_1 = \frac{\beta_1}{2} \frac{(V_R + \Delta V_{t1} - \Delta V_{t3})^2}{\left(\sqrt{\frac{n}{m} \cdot \left(1 + \frac{\Delta n}{n}\right) \frac{\left(1 + \frac{\Delta \beta_1}{\beta_1}\right)(1 + \lambda_1 V_{DS1})}{\left(1 + \frac{\Delta \beta_3}{\beta_3}\right)(1 + \lambda_3 V_{DS3})}} - 1 \right)^2} \tag{A.19}$$

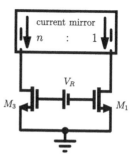

Fig. A.1 Circuit for calculation of the accuracy of the current reference

where the notation introduced in the previous section has been adopted and Δn is the error in the current mirror ratio. By using first-order approximation, i.e. by approximating the relative error in I_1 due to the error x with

$$\frac{\Delta I_1}{I_1} \approx \frac{1}{I_1} \cdot \frac{\partial \Delta I_1}{\partial x} \cdot x \tag{A.20}$$

the following expression is derived

$$\frac{\Delta I_1}{I_1} \approx 2\frac{\Delta V_{t1} - \Delta V_{t3}}{V_R} +$$

$$+ \left(\frac{n}{m} - \sqrt{\frac{n}{m}}\right)\left(\frac{\Delta n}{n} + \frac{\Delta \beta_1}{\beta_3} - \frac{\Delta \beta_3}{\beta_3} + \lambda_1 V_{DS1} - \lambda_3 V_{DS3}\right) \tag{A.21}$$

Since data about the matching of the output resistance is usually unavailable, by observing that

$$\lambda_1 V_{DS1} - \lambda_3 V_{DS3} \le \max\{\lambda_1, \lambda_3\}\,(V_{DS1} - V_{DS3}) \tag{A.22}$$

an upper bound of the standard deviation can be found:

$$\frac{\Delta I_1}{I_1} \le 2\frac{\Delta V_{t1} - \Delta V_{t3}}{V_R} + \left(\frac{n}{m} - \sqrt{\frac{n}{m}}\right)\left[\frac{\Delta n}{n} + \frac{\Delta \beta_1}{\beta_3} - \frac{\Delta \beta_3}{\beta_3} +\right.$$

$$\left. + \max\{\lambda_1, \lambda_3\}\,(V_{DS1} - V_{DS3})\right] \tag{A.23}$$

With reference to the results in the previous section, the standard deviation of the current error is then

$$\sigma_{\frac{\Delta I_1}{I_1}} \le \sigma_{V_{tu}} \frac{\sqrt{2}}{V_R}\sqrt{\frac{1}{m_1} + \frac{1}{m_3}} + \tag{A.24}$$

$$+ \sigma_{\frac{\Delta \beta_u}{\beta_u}}\frac{1}{\sqrt{2}}\left(\frac{n}{m} - \sqrt{\frac{n}{m}}\right)\sqrt{\frac{1}{m_1} + \frac{1}{m_3}} + \tag{A.25}$$

$$+ \sigma_{\Delta V_{DS}}\left(\frac{n}{m} - \sqrt{\frac{n}{m}}\right)\max\{\lambda_1, \lambda_3\} \tag{A.26}$$

$$+ \sigma_{\frac{\Delta n}{n}}\left(\frac{n}{m} - \sqrt{\frac{n}{m}}\right) \tag{A.27}$$

where $\sigma_{\Delta V_{DS}}$ and $\sigma_{\frac{\Delta n}{n}}$ are the standard deviation of $V_{DS1} - V_{DS3}$ and $\frac{\Delta n}{n}$, respectively.

Fig. A.2 Circuit for the
calculation of the effect of
leakage on the accuracy of
the current reference

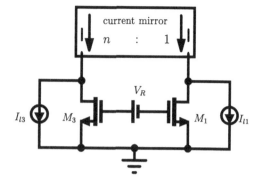

To simplify the notation, the following definitions are used:

$$\Delta V_{t1,3} = \Delta V_{t1} - \Delta V_{t3} \tag{A.28}$$

$$\frac{\Delta \beta_{1,3}}{\beta_{1,3}} = \frac{\Delta \beta_1}{\beta_3} - \frac{\Delta \beta_3}{\beta_3} \tag{A.29}$$

and (A.23) becomes

$$\frac{\Delta I_1}{I_1} \leq 2\frac{\Delta V_{t1,3}}{V_R} + \left(\frac{n}{m} - \sqrt{\frac{n}{m}}\right)\left[\frac{\Delta n}{n} + \frac{\Delta \beta_{1,3}}{\beta_{1,3}} + \max\{\lambda_1, \lambda_3\}(V_{DS1} - V_{DS3})\right] \tag{A.30}$$

A.2.2   Effect of Leakage

The effect of leakage on the current I_1 of the current reference is derived by using
the circuit of Fig. A.2, in which I_{l1} and I_{l3} are the leakage current at the drain of M_1
and M_3, respectively. The system (A.18) is modified as

$$\begin{cases} I_3 + I_{l3} = n(I_1 + I_{l1}) \\ V_{GS3} \quad = V_{GS1} + V_R \end{cases} \tag{A.31}$$

The current can be written as

$$I_1 = \frac{\beta_1}{2}\frac{V_R^2}{\left(\sqrt{\frac{n}{m}\cdot\left(1 + \frac{nI_{l1} - I_{l3}}{nI_1}\right)} - 1\right)^2} \tag{A.32}$$

and the leakage affects the current in a similar way as an error in the mirror ratio n in the previous section:

$$\frac{\Delta I_1}{I_1} \approx \left(\frac{n}{m} - \sqrt{\frac{n}{m}}\right)\frac{nI_{l1} - I_{l3}}{nI_1} \tag{A.33}$$

A.3   Matching of Current Mirrors

The attenuation (i.e. the inverse of the gain) of a current mirror (as the one in Fig. A.3) is given by

$$n = \frac{I_4}{I_2} = \frac{\dfrac{\beta_4}{2}(V_{GS} - V_{t4})^2}{\dfrac{\beta_2}{2}(V_{GS} - V_{t2})^2} \tag{A.34}$$

where the effect of finite transistor output impedance has been neglected. M_2 and M_4 are composite transistor (see Sect. A.1) constituted by n_2 and n_4 nominally equal unit elements, respectively. Proceeding as in the previous section and using a similar notation, it is possible to write

$$n = \frac{n_4}{n_2} \cdot \frac{1 + \dfrac{\Delta\beta_4}{\beta_4}}{1 + \dfrac{\Delta\beta_2}{\beta_2}} \cdot \frac{\left(1 - \dfrac{\Delta V_{t4}}{V_{GS} - V_{t4}}\right)^2}{\left(1 - \dfrac{\Delta V_{t2}}{V_{GS} - V_{t2}}\right)^2} \tag{A.35}$$

and the relative error is

$$\frac{\Delta n}{n} \approx \frac{\Delta\beta_4}{\beta_4} - \frac{\Delta\beta_2}{\beta_2} + 2\frac{\Delta V_{t2} - \Delta V_{t4}}{V_{GS} - V_{t2,0}} \tag{A.36}$$

$$\approx \frac{\Delta\beta_{2,4}}{\beta_{2,4}} + 2\frac{\Delta V_{t2,4}}{V_{GS} - V_t} \tag{A.37}$$

where it has been used that $\frac{\Delta\beta_{2,4}}{\beta_{2,4}} = \frac{\Delta\beta_4}{\beta_4} - \frac{\Delta\beta_2}{\beta_2}$, $\Delta V_{t2,4} = \Delta V_{t2} - \Delta V_{t4}$ and $V_{t2,0} = V_{t4,0} = V_t$.

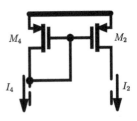

Fig. A.3 Current mirror

A.4 Behavior of MOS Transistor Mismatch Over Temperature

It has been shown in Sect. 4.4.1 that the accuracy of a trimmed and temperature-compensated time reference is given by

$$\frac{f_{comp}(T) - f_0(T_0)}{f_0(T_0)} \approx \frac{\Delta f(T)}{f_0(T)} - \frac{\Delta f(T_0)}{f_0(T_0)} \tag{A.38}$$

The previous equation shows that the accuracy is limited by the *temperature variations* of the frequency errors. Since frequency errors are also due to mismatch of MOS parameters (threshold voltage, current factor β), an estimate of their temperature behavior is needed.

The mismatch in threshold voltage (ΔV_t) and in current factor ($\Delta \beta / \beta$) are usually assumed temperature independent and their temperature coefficient is usually not modeled. In the following two sections, estimates of their temperature behavior are given based on data available in the open literature. Those estimates do not have general validity, since data from one or two particular experiments in particular technologies are used. However, a feeling for the order of magnitude of such parameters results from such analysis.

At room temperature, the mismatch of the parameters of a matched pair of transistors with length L and width W is characterized by a standard deviation given by [1]

$$\sigma_{\Delta V_t} = \frac{A_{V_t}}{\sqrt{WL}} \tag{A.39}$$

$$\sigma_{\Delta \beta / \beta} = \frac{A_\beta}{\sqrt{WL}} \tag{A.40}$$

where the matching constants for modern CMOS technologies are in the order of $A_{V_t} \sim 5\text{–}10\,\text{mV} \cdot \mu\text{m}$ and $A_\beta \sim 1\text{–}2\%\cdot\mu\text{m}$ [2].

A.4.1 Threshold Voltage

In [3], the offset voltage of a CMOS differential amplifier (implemented in a $0.5\text{-}\mu\text{m}$ BiCMOS technology) is approximated as

$$V_{os} = \Delta V_t + \frac{I_D}{g_m} \frac{\Delta \beta}{\beta} \tag{A.41}$$

considering that the offset is dominated by the mismatch of the matched input pair. After individual room-temperature trim of both ΔV_t and $\frac{\Delta \beta}{\beta}$, a residual temperature coefficient of the offset of $0.33\,\mu\text{V/°C}$ (3σ value) has been measured over the temperature range from -40 to $125°\text{C}$.

An upper bound to the temperature coefficient of the threshold mismatch is given by assuming it as the only contributor to the final temperature drift. This would correspond to a maximum temperature coefficient for ΔV_t of $0.33\,\mu V/°C$ and to a variation in the order of less than $100\,\mu V$ over the military temperature range after room temperature trim.

A.4.2   Current Factor (β)

With the approach of the previous section on the data from [3], the following bound can be derived

$$\frac{\Delta\beta}{\beta} < \frac{g_m}{I_D}V_{os} = \frac{qV_{os}}{nkT} < \frac{qV_{os}}{kT} \tag{A.42}$$

where it has been used the fact that the transistors are biased in weak inversion and that the subthreshold slope factor $n > 1$. This would give a maximum temperature coefficient for $\frac{\Delta\beta}{\beta}$ of $16\,\mathrm{ppm/°C}$ and to a variation in the order of less than 0.2% over the military temperature range after room temperature trim.

Data on the temperature dependence of the current factor mismatch can also be found in [4]. The mismatch of the drain current of a matched pair (implemented in a 65-nm CMOS technology with $W = L = 1\,\mu m$) after trimming at room temperature is reported. Since the matched pair is biased with a gate-source voltage equal to the nominal supply for that technology, the effect of the threshold voltage mismatch is negligible, making an estimation of the β mismatch more accurate. For the reported data, the variation over a temperature range of $100°C$ is less than 0.5%, which is in the same order of magnitude of the previous estimation.

A.5   Charge of a MOS Capacitor

To quantify the effect of the MOS capacitance on its charge time and, consequently, on the oscillator's output frequency, the model of Fig. A.4 is adopted. The charge (or discharge) time T of the capacitor between the two threshold voltages V_1 and V_2

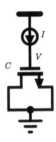

Fig. A.4 Circuit illustrating the charge of a MOS capacitor

$(V_2 > V_1)$ with a constant current I, can be found with few passages, starting from the conservation of charge Q at the gate:

$$I = \frac{dQ}{dt} \tag{A.43}$$

$$\int_{t_1}^{t_2} I \, dt = \int_{Q_1}^{Q_2} dQ \tag{A.44}$$

$$I \cdot T = \int_{V_1}^{V_2} C \, dV \tag{A.45}$$

$$T = \frac{1}{I} \int_{V_1}^{V_2} C \, d = \frac{Q_2 - Q_1}{I} V \tag{A.46}$$

where $C = \frac{\partial Q}{\partial V}$ is the small-signal capacitance $Q_{1,2}$ is the charge on the top plate of C when the voltage applied across the capacitor is $V_{1,2}$, respectively.

If the small-signal capacitance were constant and equal to the oxide capacitance $C = WLC_{ox}$ for applied voltages between V_1 and V_2, the ideal charging time would be

$$T_{ideal} = \frac{1}{I} \int_{V_1}^{V_2} C \, dV = WLC_{ox} \frac{V_2 - V_1}{I} \tag{A.47}$$

To derive the effective charging time, the following approximate expressions from [5], valid for a MOS structure in strong inversion, are adopted:

$$Q = WL\sqrt{2q\epsilon_s N_A}\sqrt{\psi_s + \phi_t \exp\left(\frac{\psi_s - 2\phi_F}{\phi_t}\right)} \tag{A.48}$$

$$V = \phi_{MS} + \psi_s + \frac{\sqrt{2q\epsilon_s N_A}}{C_{ox}}\sqrt{\psi_s + \phi_t \exp\left(\frac{\psi_s - 2\phi_F}{\phi_t}\right)} \tag{A.49}$$

$$C_{ox} = \frac{\epsilon_{ox}}{t_{ox}} \tag{A.50}$$

where Q is the charge on the capacitor for an applied voltage V, q the electron charge, ϵ_s the dielectric constant of silicon, N_A the doping concentration in the substrate, ψ_s the potential in the MOS channel, ϕ_F the Fermi level of the silicon substrate, $\phi_t = kT/q$ the thermal voltage, ϕ_{MS} the difference in working function between the silicon substrate and the gate, ϵ_{ox} the dielectric constant of silicon dioxide and t_{ox} the gate oxide thickness. By combining the previous equations, the relative error committed in approximating T with T_{ideal} is

$$e = \frac{T_{ideal} - T}{T_{ideal}} = \frac{\psi_s(V_2) - \psi_s(V_1)}{V_2 - V_1} \tag{A.51}$$

Fig. A.5 MOS-charge-time
error for different
combinations of process
parameters
$(N_A = 10^{17} - 10^{18}\,\text{cm}^{-3},$
$t_{ox} = 2 - 30\,\text{nm})$ for
$V_2 - V_1 = 0.4\,\text{V}$, according to
the presented model and the
simulated error for the
thick-oxide transistors of the
65-nm process described in
Chap. 4

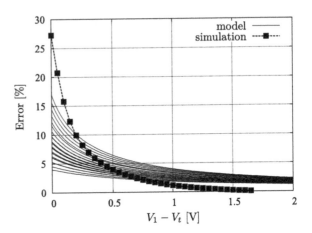

Fig. A.6 MOS-charge-time
error for different
combinations of process
parameters for $V_1 - V_t = 1\,\text{V}$
and $V_2 - V_1 = 0.4\,\text{V}$,
according to the presented
model and the simulated error
for the thick-oxide transistors
of the 65-nm process
described in Chap. 4

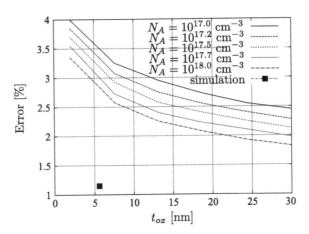

Since an analytical expression for ψ_s can not be derived, the error e has been
numerically computed for values of N_A and t_{ox} typical of modern CMOS processes.
In order to get the results comparable for the different process parameters, the
following expression for the threshold voltage has been used [5]:

$$V_t = \phi_{MS} + \phi_0 + \frac{\sqrt{2q\epsilon_s N_A}}{C_{ox}}\sqrt{\phi_0} \qquad (A.52)$$

with $\phi_0 = 2\phi_F + 6\phi_t$. The results of the computation are shown in Figs. A.5
and A.6. To assess the effectiveness of the model, the charge time error for a specific
process (i.e. the 65-nm CMOS process described in Chap. 4) has been simulated and
plotted in the figures. The simulation and the analytical model brings to results in
the same order of magnitude. It is then clear that a difference of at least 1% can be
expected in the real case with respect to the simple model that assumes $C = C_{ox}$.

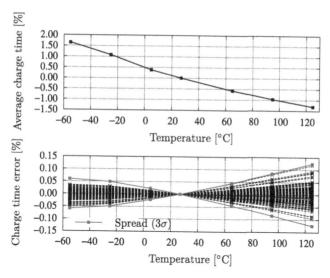

Fig. A.7 Simulated MOS-charge-time (average and error with respect to the average) for the thick-oxide transistors of the 65-nm process described in Chap. 4 ($V_1 = 0.8\,V$, $V_1 = 1.2\,V$, $V_t \approx 0.45\,V$ at room temperature); all values are normalized to the value at room temperature

However, as discussed in Sect. 4.4.1, a trimmed reference is only affected by the temperature variations of the errors and not by their absolute value. Thus, if the temperature variations of e are smaller than its absolute value, the effect of a real MOS capacitor on the reference accuracy would be less than 1%.

To prove the latter point, Monte Carlo simulations of the MOS capacitor charging in the 65-nm process previously cited have been run and the results are reported in Fig. A.7. Those results prove that, if the temperature compensation of the reference takes also into consideration the average temperature behavior of the charge time, the accuracy is only affected by the spread in the temperature behavior for the charge time, which is shown to be smaller of 0.15% over the military temperature range. Note that in this simulation, and also in the implementation of Chap. 4, temperature-independent voltages are used for V_1 and V_2. Even if choosing voltages tracking the temperature behavior of the threshold voltage could lower the temperature coefficient of the charge time, it is more practical the use of constant voltages.

References

1. Pelgrom M, Duinmaijer A, Welbers A (1989) Matching properties of MOS transistors. IEEE J Solid State Circ 24(5):1433–1439
2. Kinget PR (2005) Device mismatch and tradeoffs in the design of analog circuits. IEEE J Solid State Circ 40(6):1212–1224

3. Bolatkale M, Pertijs MAP, Kindt WJ, Huijsing JH, Makinwa KAA (2008) A BiCMOS operational amplifier achieving $0.33\mu V/^{\circ}C$ offset drift using room-temperature trimming. In: ISSCC Dig. of Tech. Papers, pp 76–77
4. Andricciola P, Tuinhout HP (2009) The temperature dependence of mismatch in deep-submicrometer bulk MOSFETs. IEEE Electron Dev Lett 30(6):690–692
5. Tsividis Y (2003) Operation and modeling of the MOS transistor, 2nd edn. Oxford University Press, New York

Appendix B
Analysis of the Spread of the Mobility-Based Time Reference

B.1 Model for the Frequency Error

Following the reasoning in Sect. 4.4.5, the output frequency of the mobility-based time reference can be expressed as:

$$f(T) = k \cdot (\mu_0 + \Delta\mu_0) \left(\frac{T}{T_0}\right)^{\alpha_\mu + \Delta\alpha} \tag{B.1}$$

where k is a proportionality constant and $\Delta\mu_0$ and $\Delta\alpha$ are variations due to the process spread. The frequency error Δf [see (4.2)] can be approximated as

$$\Delta f(T) \approx \left.\frac{\partial f}{\partial \mu_0}\right|_{\Delta\mu_0=0} \cdot \Delta\mu_0 + \left.\frac{\partial f}{\partial \alpha_\mu}\right|_{\Delta\alpha=0} \cdot \Delta\alpha \tag{B.2}$$

$$= k \cdot \Delta\mu_0 \left(\frac{T}{T_0}\right)^{\alpha_\mu} - k \cdot \Delta\alpha \cdot \mu_0 \log\left(\frac{T}{T_0}\right)\left(\frac{T}{T_0}\right)^{\alpha_\mu} \tag{B.3}$$

$$= k \cdot \Delta\mu_0 \cdot g_\mu(T) - k \cdot \Delta\alpha \cdot \mu_0 \cdot g_\alpha(T) \tag{B.4}$$

where

$$g_\mu(T) = \left(\frac{T}{T_0}\right)^{\alpha_\mu} \tag{B.5}$$

$$g_\alpha(T) = \log\left(\frac{T}{T_0}\right)\left(\frac{T}{T_0}\right)^{\alpha_\mu} \tag{B.6}$$

F. Sebastiano et al., *Mobility-based Time References for Wireless Sensor Networks*, Analog Circuits and Signal Processing, DOI 10.1007/978-1-4614-3483-2, © Springer Science+Business Media New York 2013

Assuming a correlation c among the random variables $\Delta\mu_0$ and $\Delta\alpha \cdot \mu_0$, such that

$$\Delta\alpha \cdot \mu_0 = c \cdot \Delta\mu_0 + \epsilon \tag{B.7}$$

with $\Delta\mu_0$ and ϵ independent random variables, the variance of Δf can be computed as

$$\sigma_{\Delta f}^2 = \left[g_\mu(T) + c \cdot g_\alpha(T) \right]^2 \sigma_{\Delta\mu_0}^2 + g_\alpha^2(T)\sigma_\epsilon^2 \tag{B.8}$$

where $\sigma_{\Delta\mu_0}$ and σ_ϵ are the standard deviations of $\Delta\mu_0$ and ϵ, respectively.

The standard deviation of error of the trimmed and temperature-compensated reference can be derived directly from (4.25):

$$\sigma_{trim} = \sigma_{\Delta\alpha} \log\left(\frac{T}{T_0}\right) \tag{B.9}$$

where $\sigma_{\Delta\alpha}$ is the standard deviations of $\Delta\alpha$.

B.2   Analysis of Experimental Data

Data from the two 65-nm CMOS batches presented in Chaps. 4 and 5 were analyzed using the model presented in the previous section. For each sample, the measured error Δf was least-square fit using (B.4), so that the two parameters $k \cdot \Delta\mu_0$ and $k \cdot \Delta\alpha\mu_0$ were extracted for each sample. Figure B.1 shows the result of the fitting for the first batch, while Fig. B.2 reports the computed parameters for both batches. Since a limited number of samples were available, it is difficult to draw conclusions

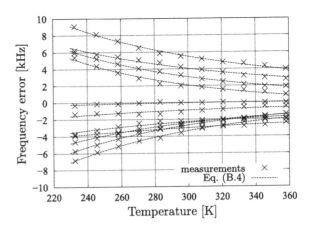

Fig. B.1 Experimental data and least-square fit of the model for the first 65-nm CMOS batch

Fig. B.2 Coefficients of the least-square fit of (B.4)

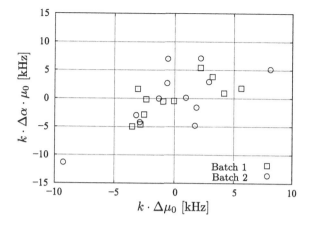

Fig. B.3 Comparison between measured standard deviation of Δf and the model of (B.8)

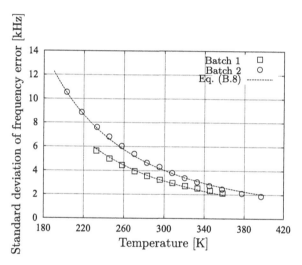

Fig. B.4 Standard deviation of the error of the trimmed and temperature-compensated time reference

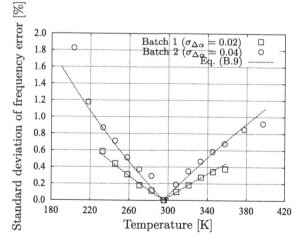

about the correlation of the two coefficients. However, some insight can be gained by observing the standard deviation of Δf shown in Fig. B.3. The fitting of (B.8) shown in the figure was obtained by assuming $c = 0.9$. The standard deviation of $k \cdot \Delta \mu_0$ and $k \cdot \Delta \alpha \cdot \mu_0$ were computed from the data in Fig. B.2 and σ_ϵ was derived using (B.7). The good agreement between the experimental data and the model hints to some correlation between $\Delta \mu_0$ and $\Delta \alpha \cdot \mu_0$.

Regarding the trimmed and temperature-compensated reference, its standard deviation is shown in Fig. B.4 together with the fit of (B.9). Note that for the second batch, the point at lowest temperature has not been considered for the fitting of the model.

Glossary

Symbols

α	Scale factor for ΔV_{be} in the temperature sensor	[-]
α_μ	Exponent of the temperature behavior of the electron mobility	[-]
α_{tot}	Temperature coefficient of composite resistor	[1/K]
α_n	Temperature coefficient of the resistor with negative temperature coefficient	[1/K]
α_p	Temperature coefficient of the resistor with positive temperature coefficient	[1/K]
β	When used for a bipolar transistor: current gain ($\beta = \frac{I_C}{I_B}$)	[A/V^2]
β	When used for a MOS transistor: β factor of a MOS transistor ($\beta = \mu_n C_{ox} \frac{W}{L}$)	[A/V^2]
$\Delta\alpha$	Variation of the exponent of the temperature behavior of the electron mobility due to process spread	[-]
$\Delta\mu_0$	Variation of the electron mobility at temperature T_0 due to process spread	[m^2/(V·s)]
Δf	Error in oscillator frequency due to process spread	[Hz]
Δf_{comp}	Error in the oscillator frequency after trimming and temperature compensation	[Hz]
Δf_{osc}	Difference between nominal frequency and local oscillator frequency	[Hz]
Δf_{syn}	Frequency error at the receiver	[Hz]
Δf_{TX}	Allowed deviation of the transmitting frequency	[Hz]
Δt_{syn}	Timing error at the receiver	[s]

F. Sebastiano et al., *Mobility-based Time References for Wireless Sensor Networks*,
Analog Circuits and Signal Processing, DOI 10.1007/978-1-4614-3483-2,
© Springer Science+Business Media New York 2013

ΔV_{be}	Voltage difference between the base and emitter voltage of two bipolar transistors	[V]
θ	Systematic phase variation	[rad]
μ	Average of the output bitstream of the temperature sensor	[-]
μ_0	Electron mobility at temperature T_0	$[m^2/(V \cdot s)]$
μ_n	Electron mobility	$[m^2/(V \cdot s)]$
μ_p	Hole mobility	$[m^2/(V \cdot s)]$
μ_{PTAT}	Compensated output (proportional to absolute temperature) of the temperature sensor	[-]
ρ	Resistivity	$[\Omega \cdot m]$
$\sigma_j(\tau)$	Jitter accumulated after a time τ	[s]
σ_y	Allan variance	[s]
ϕ	Random phase variations	[rad]
ψ	Instantaneous phase	[rad]
ω	Angular frequency ($\omega = 2\pi f$)	[rad/s]
a_{time}	Relative inaccuracy of the time reference	[-]
B_{limit}	Width of the transmitted spectrum at -20 dBm/MHz	[Hz]
B_n	Receiver noise bandwidth	[Hz]
BER	Bit Error Rate	[-]
BR	Synchronization beaconing rate	[Hz]
C	Capacitance	[F]
c	Light speed ($c = 2.998 \cdot 10^8$ m/s)	[m/s]
C_{ox}	MOS oxide capacitance per unit area	$[F/m^2]$
CF	Crest factor	[-]
$CF_{TX,max}$	Transmitter limit for the crest factor	[-]
d	Maximum distance between two nodes	[m]
DC_{clk}	Duty-cycle of the time reference	[-]
DC_{rx}	Duty-cycle of the receiving section of the main radio	[-]
DC_{tx}	Duty-cycle of the transmitting section of the main radio	[-]
DC_{wu}	Duty-cycle of the wake-up radio	[-]
DR	Data rate	[Hz]
E_p	Energy detected by the correlation receiver	[-]
f	Frequency	[Hz]
f_0	Nominal oscillator frequency	[Hz]
f_{comp}	Oscillator frequency after trimming and temperature compensation	[Hz]
f_{osc}	Frequency of the input signal to the temperature-compensation circuit	[Hz]

f_{out}	Frequency of the output of the divider used for temperature compensation	[Hz]
f_{RF}	Output frequency of the PLL used for temperature compensation	[Hz]
f_{trim}	Oscillator frequency after trimming	[Hz]
g_m	Transconductance	[S]
I_B	Base current of a bipolar transistor	[A]
I_C	Collector current of a bipolar transistor	[A]
I_D	Drain current of a MOS transistor	[A]
I_S	Saturation current of a bipolar transistor	[A]
IL_{syn}	Implementation loss for the receiver synchronization	[-]
$J_{rel}(\tau)$	Relative jitter accumulated after a time τ	[-]
k	Boltzmann constant ($k = 1.381 \cdot 10^{-23}$ J/K)	[J/K]
$\mathscr{L}(f)$	Phase noise at a frequency offset f from the carrier	[dBc/Hz]
L	Inductance	[H]
L	Length (of resistors or transistors)	[m]
m	Number of frame periods used by the wake-up radio to take a decision	[-]
n	Current gain of the current mirror used to bias the bipolar core of the temperature sensor	[-]
N_{div}	Division factor of the frequency divider used for temperature compensation of the oscillator frequency	[-]
n_{fa}	Average number of false alarms per timeslot	[-]
n_{md}	Average number of missed detections per packet	[-]
N_{mul}	Multiplication factor of the PLL used for temperature compensation of the oscillator frequency	[-]
n_{nodes}	Number of nodes in the network	[-]
n_{pkt}	Number of packets to be received in one timeslot	[-]
N_{pl}	Packet payload (bits in a packet)	[-]
n_{PL}	Path loss coefficient	[-]
N_{pr}	Number of bits in the packet preamble	[-]
N_{res}	Resolution of N_{div} and N_{mul}	[-]
n_{sub}	MOS subthreshold slope factor	[-]
n_{th}	Threshold for the number of detected pulses to issue a wake-up call	[-]
n_p	Number of pulses per bit in the PPM	[-]
NF	Noise figure	[-]
p	Pulse shape of the modulation	[-]
P_{avg}	Average transmitted power	[W]
P_{clk}	Peak power consumption of the time reference	[W]
$\mathbb{P}_{fa}$	Probability that a decision results in a false alarm	[-]

P_{fp}	False pulse probability	[-]
P_{int}	Interferer power	[W]
$\mathbb{P}_{md}$	Probability that a decision results in a missed detection	[-]
p_{mp}	Missed pulse probability	[-]
$P_{peak,limit}$	Transmitter limit for the peak power of the transmitted signal	[W]
P_{peak}	Peak power of the transmitted signal	[W]
P_{rx}	Peak power consumption of the receiving section of the main radio	[W]
P_{tx}	Peak power consumption of the transmitting section of the main radio	[W]
P_{wu}	Peak power consumption of the wake-up radio	[W]
PL	Path loss	[-]
PR	Packet rate	[Hz]
PRF	Pulse repetition frequency in the PPM	[Hz]
q	Electron charge ($q = 1.6 \cdot 10^{-19}$ C)	[C]
Q_C	Capacitor quality factor	[-]
Q_L	Inductor quality factor	[-]
$\mathbf{R}\{\cdot\}$	Real part of a complex number	[-]
R_{end}	Contact resistance	[Ω]
R_{n0}	Extrapolated resistance at 0 K of the resistor with negative temperature coefficient	[Ω]
R_{p0}	Extrapolated resistance at 0 K of the resistor with positive temperature coefficient	[Ω]
R_{sh}	Sheet resistance	[Ω]
r_0	Reference distance for the path loss calculation ($r_0 = 1$ m)	[m]
R_B	Parasitic resistance in series with the base of a bipolar transistor	[Ω]
R_C	Capacitor series resistance	[Ω]
R_E	Parasitic resistance in series with the emitter of a bipolar transistor	[Ω]
R_L	Inductor series resistance	[Ω]
S	Receiver sensitivity	[W]
S_ϕ	Single-sided power spectral density of random phase variations ϕ	[1/Hz]
S_x^y	Sensitivity factor of parameter y due to a variation of variable x ($S_x^y = \frac{\Delta y}{y} / \frac{\Delta x}{x}$)	[-]
SNR	Signal-to-noise ratio at the input of the envelope detector	[-]
T	Absolute temperature	[K]

t	Thickness (of resistors or layers)	[m]
t	Time	[s]
T_{guard}	Guard time between timeslots	[s]
T_{pkt}	Packet transmission time	[s]
T_{ppm}	Pulse delay for bit 0 in the PPM	[s]
T_{wu}	Duration of listening timeslot of the wake-up radio	[s]
T_0	Trimming temperature	[K]
T_b	Bit duration	[s]
T_d	Time required to the wake-up radio to make a decision	[s]
T_f	Pulse repetition period of the PPM	[s]
T_p	Pulse duration in the PPM	[s]
V_{be}	Voltage difference between the base and emitter of a bipolar transistor	[V]
V_{ce}	Voltage difference between the collector and emitter of a bipolar transistor	[V]
V_{DS}	Voltage difference between drain and source of a MOS transistor	[V]
V_{GS}	Voltage difference between gate and source of a MOS transistor	[V]
V_{th}	Threshold of the comparator in the wake-up radio	[V]
V_t	Threshold voltage of a MOS transistor	[V]
W	Width (of resistors or transistors)	[m]
x_I	In-phase component of the input of the correlation receiver	[-]
x_Q	Quadrature component of the input of the correlation receiver	[-]
y_I	In-phase component of the output of the correlation receiver	[-]
y_Q	Quadrature component of the output of the correlation receiver	[-]

Index

F. Sebastiano et al., *Mobility-based Time References for Wireless Sensor Networks*,   169
Analog Circuits and Signal Processing, DOI 10.1007/978-1-4614-3483-2,
© Springer Science+Business Media New York 2013